DMV Seminar
Band 19

Springer Basel AG

Piet Groeneboom
Jon A. Wellner

Information Bounds and Nonparametric Maximum Likelihood Estimation

Springer Basel AG

Authors' addresses:

Piet Groeneboom
Delft University of Technology
Dept. of Mathematics and Computer Science
Mekelweg 4
NL–2628 CD Delft

Jon A. Wellner
Dept. of Statistics GN–22
University of Washington
Seattle, WA 98195
USA

A CIP catalogue record for this book is available from the Library of Congress, Washington D.C., USA

Deutsche Bibliothek Cataloging-in-Publication Data

Groeneboom, Piet:
Information bounds and nonparametric maximum likelihood estimation /
Piet Groeneboom ; Jon A. Wellner. – Basel ; Boston ; Berlin : Birkhäuser, 1992
  (DMV-Seminar ; Bd. 19)
  ISBN 978-3-7643-2794-1   ISBN 978-3-0348-8621-5 (eBook)
  DOI 10.1007/978-3-0348-8621-5
NE: Wellner, Jon A.:; Deutsche Mathematiker-Vereinigung: DMV-Seminar

© 1992 Springer Basel AG
Originally published by Birkhäuser Verlag Basel in 1992
ISBN 978-3-7643-2794-1

Introduction

This book contains the lecture notes for a DMV course presented by the authors at Günzburg, Germany, in September, 1990. In the course we sketched the theory of information bounds for nonparametric and semiparametric models, and developed the theory of nonparametric maximum likelihood estimation in several particular inverse problems: interval censoring and deconvolution models.

Part I, based on Jon Wellner's lectures, gives a brief sketch of information lower bound theory: Hájek's convolution theorem and extensions, useful minimax bounds for parametric problems due to Ibragimov and Has'minskii, and a recent result characterizing differentiable functionals due to van der Vaart (1991). The differentiability theorem is illustrated with the examples of interval censoring and deconvolution (which are pursued from the estimation perspective in part II). The differentiability theorem gives a way of clearly distinguishing situations in which the parameter of interest can be estimated at rate $n^{1/2}$ and situations in which this is not the case. However it says nothing about which rates to expect when the functional is *not* differentiable. Even the casual reader will notice that several models are introduced, but not pursued in any detail; many problems remain.

Part II, based on Piet Groeneboom's lectures, focuses on nonparametric maximum likelihood estimates (NPMLE's) for certain inverse problems. The first chapter deals with the interval censoring problem. In this model, one only observes the endpoints of an interval to which the variable of interest belongs, a situation quite common in medical research. The classical approach (see, e.g., Turnbull (1974)) is to use the EM-algorithm. This is shown to correspond to the so-called "self-consistency equations" for the NPMLE. These equations yield necessary but not sufficient conditions for the characterization of the NPMLE and have (so far) not been very successful in developing distribution theory. For this reason we turn to another approach, based on isotonic regression theory, which gives necessary and sufficient conditions. Moreover, it yields efficient algorithms and also leads to distribution theory, to be developed in Chapter 5.

The second chapter follows a similar path as the first chapter, but this time for deconvolution problems. Chapter 3 discusses algorithms and in Chapter 4 consistency is proved for the NPMLE in all situations discussed in the preceding chapters. Chapter 5 deals with distribution theory and discusses open problems and conjectures.

We want to thank the Gesellschaft für Mathematische Forschung, the Deutsche Mathematiker Vereinigung and in particular Professors Lerche and Witting for organizing the seminar and for all the help in making this meeting so pleasant. The first author wants to thank Ronald Geskus and Rik Lopuhaä for their comments on the manuscript and Wim Penninx for helping in translating the TEX files into one LaTeX file. The second author wants to express his thanks to Peter Sasieni, Aad van der Vaart, his three co-authors Peter J. Bickel, Chris Klaassen and Ya'acov Ritov of the book "Efficient and Adaptive Estimation for Semiparametric Models", and to Krjsten Henriksen for helping him to transfer Troff to T_EX.

We both want to thank the participants for their constructive remarks during the course. Finally we want to thank Birkhäuser Verlag for the opportunity to publish the seminar notes.

Contents

Part I

Information Bounds

1 Models, Scores, and Tangent Spaces

1.1 Introduction

In this first lecture our goal is to give a quick introduction to several key concepts for modern large-sample statistical theory, and to illustrate these concepts with several models which we will pursue in considerable detail in the remainder of the course.

The key ideas are those of *score functions* of models, and the *tangent sets* and *tangent spaces* generated by the score functions. More generally, in dealing with semiparametric models, it will be useful to deal with *score operators*.

1.2 Models $\mathcal{P}$

Consider a given sample space $\mathcal{X}$ with a σ-field of subsets $\mathcal{A}$, so that $(\mathcal{X}, \mathcal{A})$ is a measurable space. A *model* $\mathcal{P}$ is a collection of probability measures P on $(\mathcal{X}, \mathcal{A})$. Thus, for each $P \in \mathcal{P}$, $(\mathcal{X}, \mathcal{A}, P)$ is a probability space (some authors call $\{(\mathcal{X}, \mathcal{A}, P) : P \in \mathcal{P}\}$ an *experiment*).

Often the statistical problem can be phrased as follows. If $X_1, \ldots, X_n$ are independent and identically distributed observations from $P \in \mathcal{P}$, estimate P or some functional $\nu(P)$.

Example 1.1. (Parametric models). If

$$\mathcal{P} = \{P_\theta : \theta \in \Theta\}, \text{ where } \Theta \subset \mathbb{R}^k$$

for some k, then $\mathcal{P}$ is a *parametric model* . For example if $(\mathcal{X}, \mathcal{A}) = (\mathbb{R}^+, \mathcal{B})$, the nonnegative real numbers with its usual Borel σ-field, then the *Weibull* family $\mathcal{P}$ is the parametric model with densities

$$p_\theta(x) = \frac{\beta}{\alpha}(\frac{x}{\alpha})^{\beta-1} \exp(-(\frac{x}{\alpha})^\beta) 1_{[0,\infty)}(x)$$

with respect to Lebesgue measure where $\theta \equiv (\alpha, \beta) \in (0, \infty) \times (0, \infty) \subset \mathbb{R}^2$.

Example 1.2. (Nonparametric models). Let μ be a fixed given σ-finite measure on $(\mathcal{X}, \mathcal{A})$. If

$$\mathcal{P} = \{\text{all } P \text{ on } (\mathcal{X}, \mathcal{A})\} \equiv \mathcal{M},$$

or if

$$\mathcal{P} = \{\mathrm{P} \in \mathcal{M} : P << \mu \text{ with density } p = \frac{dP}{d\mu}\} \equiv \mathcal{M}_\mu,$$

then $\mathcal{P}$ is often referred to as a *nonparametric model* . For example, if $(\mathcal{X}, \mathcal{A}) = (\mathbb{R}^+, \mathcal{B})$ as in example 1.1, and $\mathcal{P} = \mathcal{M}_\lambda$ where λ is Lebesgue measure, then

$\mathcal{P}$ is simply the collection of all probability distributions on $\mathbb{R}^+$ with Lebesgue densities. It is also common practice to refer to models with monotonicity or smoothness restrictions on the densities (or hazard rates or regression functions or ...) as "nonparametric." For example, the family

$$\mathcal{M}_{\lambda,\downarrow} \equiv \{P \in \mathcal{M}_\lambda : p \text{ is nonincreasing}\}$$

on $(\mathbb{R}^+, \mathcal{B})$ is a "nonparametric model."

Many problems in statistics involve *missing data* : Suppose that $X^0 \sim Q$ on $(\mathcal{X}^0, \mathcal{B})$, but we observe only $X \equiv T(X^0) \sim P \equiv QT^{-1}$ on $(\mathcal{X}, \mathcal{A})$. Several of the following examples can be thought of in this way.

Example 1.3. (Interval censoring). Suppose that $X^0 = (Z, U)$ in $[0, \infty) \times [0, \infty)$ where $Z \sim F$ and $U \sim G$ are independent, so Q is the product measure $F \times G$. We observe $X = T(X^0) \equiv (U, 1_{[Z \le U]}) \equiv (U, \delta)$. Then the model is

$$\mathcal{P} = \{P = QT^{-1} : Q = F \times G, F \in \mathcal{M}, G \in \mathcal{M}_\lambda\}.$$

Here $P \equiv P_{F,G}$ has density

$$p_{F,G}(x) = p_{F,G}(u, \delta) = F(u)^\delta (1 - F(u))^{1-\delta} g(u)$$

with respect to $\mu \equiv \lambda \times$ (counting measure) on $\mathcal{X} \equiv [0, \infty) \times \{0, 1\}$.

Example 1.4. (Convolution models). Suppose that $X^0 = (Z, W)$ in $\mathbb{R}^2$ where $Z \sim F$ is unknown, $W \sim G \in \mathcal{M}_\lambda$ is a fixed distribution, and Z, W are independent (so Q is again the product measure $F \times G$). We observe $X = T(X^0) \equiv Z + W$. Then the model is

$$\mathcal{P} = \{P = QT^{-1} : Q = F \times G, F \in \mathcal{M}\}.$$

Here $P = P_F$ has Lebesgue density

$$p_F(x) = \int g(x - z) dF(z)$$

where $g = dG/d\lambda$ is the Lebesgue density of G. This is a special case of the more general class of mixture models; see BKRW (1991) sections 4.5 and 6.5.

Example 1.5. (Multiplicative censoring and uncontaminated; or the Vardi-Zhang model). Let $0 < p < 1$ be fixed, and let δ_1 denote the measure with unit mass at 1. Suppose that $X^0 = (Z, W)$ where $Z \sim F$ on $\mathbb{R}^+$ and F is unknown, $W \sim G \equiv p\delta_1 + (1-p)Uniform(0, 1)$, and Z and W are independent, so again Q is the product measure $F \times (p\delta_1 + (1-p)U(0,1))$. Then we observe $X = T(X^0) \equiv (ZW, 1_{[W=1]}) \equiv (T, \delta)$. Thus $T = Z$ with probability p and $T = ZW1_{[W<1]}$ with probability $1-p$. If we assume that $F \in \mathcal{M}_\lambda$, then the model is

$$\mathcal{P} = \{P = QT^{-1} : Q = F \times G, \ F \in \mathcal{M}_\lambda\}.$$

If $f = dF/d\lambda$ is the Lebesgue density of F, then $P = P_F$ has density

$$p_F(x) = p_F(t, \delta) = (pf(t))^\delta (\bar{p} \int_{(t,\infty)} \frac{1}{z} dF(z))^{1-\delta} \qquad (1.1)$$

with respect to $\lambda \times$ (counting measure) on $\mathcal{X} \equiv \mathbb{R}^+ \times \{0,1\}$. If $p = 0$ so that $W \sim Uniform(0,1)$, then it turns out that this model $\mathcal{P}$ is precisely the collection $\mathcal{M}_{\lambda,\downarrow}$ of distributions on $\mathbb{R}^+$ with nonincreasing densities introduced above: a distribution has nondecreasing density if and only if its density can be written as a scale mixture of Uniform$(0,1)$, and this is just the second factor in (1.1).

This model has been studied by Vardi (1990) and by Vardi and Zhang (1990). It is a particular case of a model explored by Has'minski and Ibragimov (1983).

Example 1.6. (Interval censoring with two points; case II). Suppose that $X^0 = (Z, U, V)$ in $\mathbb{R}^+ \times (\mathbb{R}^+)^2$ where $Z \sim F$ and $(U, V) \sim H$ are independent and $U < V$ a.s.. Thus Q is the product measure $F \times H$. We observe $X = T(X^0) = (U, V, 1_{[Z \leq U]}, 1_{[Z \leq V]}) \equiv (U, V, \delta, \gamma)$. Thus the model is

$$\mathcal{P} = \{P = QT^{-1} : Q = F \times H, F \in \mathcal{M}_1, H \in \mathcal{M}_2\}.$$

Here $P = P_{F,H}$ has density

$$
\begin{aligned}
p_{F,H}(x) &= p_{F,H}(u, v, \delta, \gamma) \\
&= F^\delta(u)(F(v) - F(u))^\gamma (1 - F(v))^{1-\delta-\gamma} h(u, v)
\end{aligned}
$$

with respect to $\mu \equiv \lambda \times$ (counting measure) on $\mathcal{X} = (\mathbb{R}^+)^2 \times \{0,1\}^2$.

Example 1.7. (Double censoring). This model is very closely related to examples 1.3 and 1.6; the main difference between it and example 1.6 is that now we actually observe the survival time Z when it falls between U and V. Thus we have $X^0 = (Z, U, V)$ as in example 1.6, but now

$$X \equiv T(X^0) \equiv ((Z \wedge U) \vee V, 1_{[Z \leq U]}, 1_{[U < Z \leq V]}) \equiv (Y, \delta, \gamma).$$

Thus we take the model to be

$$\mathcal{P} = \{P = QT^{-1} : Q = F \times H, \ F \in \mathcal{M}_1, \ H \in \mathcal{M}_2\},$$

and now $P = P_{F,H} \in \mathcal{P}$ has density

$$
\begin{aligned}
p_{F,H}(x) &= p(y, \delta, \gamma) \\
&= \{M(y)f(y)\}^\gamma \{F(y)h_U(y)\}^\delta \{(1 - F(y))h_V(y)\}^{1-\gamma-\delta}
\end{aligned}
$$

where $M(y) \equiv P(U < y \leq V)$ and h_U and h_V are the marginal densities of U and V.

Nonparametric maximum likelihood estimates of F in this model have been studied by Chang and Yang (1970) and Chang (1990).

1.3 Scores: Differentiability of the Model

For "classical" regular parametric models $\mathcal{P}$, the *scores* (or *score functions*) play an important role. If $\mathcal{P} \subset \mathcal{M}_\mu$ so that the distributions $P_\theta \in \mathcal{P}$ have μ-densities p_θ, then the score functions $\dot{\mathbf{l}}_i$ are classically defined in terms of pointwise derivatives at θ (for each fixed x):

$$\dot{\mathbf{l}}_i(x) = \frac{\partial}{\partial t_i} \log p_t(x)|_{t=\theta}, \;\; i = 1, \ldots, k. \tag{1.2}$$

Then the (row) vector of scores is: $\dot{\mathbf{l}}_\theta \equiv (\dot{\mathbf{l}}_1, \ldots, \dot{\mathbf{l}}_k)$, and (assuming $E_\theta |\dot{\mathbf{l}}_i|^2 < \infty$, $i = 1, \ldots, k$) the *Fisher information matrix* $I(\theta)$ is defined by

$$I(\theta) \equiv E_\theta \dot{\mathbf{l}}_\theta^T(X) \dot{\mathbf{l}}_\theta(X). \tag{1.3}$$

It turns out the all the essential features for the theory go through, and many more models can be treated, if the pointwise derivatives in (1.2) are replaced by derivatives in $L_2(\mu)$ as follows: Note that $\sqrt{p_\theta} \in L_2(\mu)$ since $\int (\sqrt{p_\theta})^2 d\mu = \int p_\theta d\mu = 1$. Thus it makes sense to say that a parametric model $\mathcal{P} = \{P_\theta : \theta \in \Theta \subset \mathbb{R}^k\}$ is *Hellinger differentiable* at P_θ with derivative $\dot{\mathbf{l}}_\theta \in (L_2(P_\theta))^k$ if

$$\left\{ \int [\sqrt{p_{\theta+h}} - \sqrt{p_\theta} - \frac{1}{2} h^T \dot{\mathbf{l}}_\theta \sqrt{p_\theta}]^2 d\mu \right\}^{1/2} = o(|h|) \tag{1.4}$$

as $|h| \to 0$. In this case the *tangent set* $\dot{\mathcal{P}}^0$ at $P_\theta \in \mathcal{P}$ is $\{\dot{\mathbf{l}}_1, \ldots, \dot{\mathbf{l}}_k\} \subset L_2(P_\theta)$, and the *tangent space* $\dot{\mathcal{P}}$ is the closed (finite-dimensional in this case) subspace of $L_2(P_\theta)$ spanned by $\dot{\mathcal{P}}^0$. Note that since score functions always have zero means, $E_\theta \dot{\mathbf{l}}_i = 0$, $i = 1, \ldots, k$, $\dot{\mathcal{P}}$ is a subspace of $L_2^0(P_\theta) \equiv \{h \in L_2(P_\theta) : E_\theta h = 0\}$.

Example 1.1, continued. For the Weibull family $\mathcal{P}$, the model is Hellinger differentiable at every $P_\theta \in \mathcal{P}$ and the scores are

$$\dot{\mathbf{l}}_\alpha(x) \;\; = \;\; \frac{\beta}{\alpha}\{(\frac{x}{\alpha})^\beta - 1\}$$

$$\dot{\mathbf{l}}_\beta(x) \;\; = \;\; \frac{1}{\beta} - \frac{1}{\beta} \log\{(\frac{x}{\alpha})^\beta\}\{(\frac{x}{\alpha})^\beta - 1\}.$$

Thus $\dot{\mathcal{P}}$ is the two-dimensional subspace of $L_2(P_\theta)$ spanned by $\dot{\mathbf{l}}_\alpha$ and $\dot{\mathbf{l}}_\beta$, and the Fisher information matrix is:

$$I(\theta) = E(\dot{\mathbf{l}}_\theta^T(X)\dot{\mathbf{l}}_\theta(X)) = \begin{pmatrix} \dfrac{\beta^2}{\alpha^2} & \dfrac{a}{\alpha} \\[2ex] \dfrac{a}{\alpha} & \dfrac{b^2}{\beta^2} \end{pmatrix}$$

where, with $Y \sim$ exponential(1), and $\gamma \equiv .577\ldots$ Euler's constant,

$$a = -E\{(Y-1)^2 \log(Y)\} = -(1-\gamma),$$

and

$$b^2 = E\{(Y-1)\log(Y) - 1\}^2 = \frac{\pi^2}{6} + (1-\gamma)^2.$$

Note that $\det(I(\theta)) = (b^2 - a^2)\alpha^{-2} = (\pi^2/6)\alpha^{-2} > 0$, so $I(\theta)$ is nonsingular.

1.4 Tangent Sets and Tangent Spaces

For nonparametric or semiparametric models $\mathcal{P}$, we consider all Hellinger differentiable, one-dimensional parametric submodels $\mathcal{P}_0 \subset \mathcal{P}$, and the resulting collection of tangents $\{\dot{\mathbf{l}}(\cdot\,;\mathcal{P}_0) : \mathcal{P}_0 \subset \mathcal{P}\} \equiv \dot{\mathcal{P}}^0$. This is the *tangent set* for the model at the "point" $P \in \mathcal{P}$. Then the *tangent space* $\dot{\mathcal{P}} \equiv \dot{\mathcal{P}}(P)$ at $P \in \mathcal{P}$ is the closure of the linear span of $\dot{\mathcal{P}}^0$: i.e.

$$\dot{\mathcal{P}} = \overline{lin(\dot{\mathcal{P}}^0)}.$$

In the case of a nonparametric model $\mathcal{P}$, it is usually the case that $\dot{\mathcal{P}} = L_2^0(P)$; i.e. all mean-zero $L_2^0(P)$ functions h can be realized arbitrarily closely by the linear span of tangents of one-dimensional parametric submodels.

Example 1.2, continued. Fix $P_0 \in \mathcal{M}_\mu$ with μ-density p_0. For $h \in L_2^0(P_0)$ and $M > 0$, let $h_M \equiv h1_{[|h|\leq M]}$, and $h_M^0 \equiv h_M - \int h_M dP_0$. Then the one-parameter family

$$p_\theta(x) = \exp(\theta h_M^0(x) - b(\theta))p_0(x),$$

where

$$b(\theta) = \log\left\{\int \exp(\theta h_M^0(x))dP_0(x)\right\}$$

(note that the integral always exists since h_M^0 is bounded) is a one-parameter exponential family with score function (or tangent) at $\theta = 0$ given by

$$\dot{\mathbf{l}}_\theta(x) = h_M^0(x)$$

since $\dot{b}(0) = \int h_M^0(x)dP_0(x) = 0$. Since h_M^0 is arbitrarily close to h in $L_2(P_0)$ (for large M), it follows that $\dot{\mathcal{P}} = L_2^0(P_0)$.

Between the two extremes of parametric models with finite-dimensional tangent spaces and nonparametric models with tangent spaces $\dot{\mathcal{P}}$ equal to all of $L_2^0(P)$, there are many models with tangent spaces $\dot{\mathcal{P}}$ which are infinite-dimensional but *not* all of $L_2^0(P)$. In this case we say that the model $\mathcal{P}$ is a *semiparametric model*. More formally:

Definition 1.1. If $\mathcal{P}$ has tangent space $\dot{\mathcal{P}}(P)$ which is not finite-dimensional and is a proper subset of $L_2^0(P)$, then we say that *$\mathcal{P}$ is a semiparametric model at P (or simply that $\mathcal{P}$ is semiparametric)*.

1.5 Score Operators

For many semiparametric models, and especially models involving missing data, it is useful to work with score operators. In the case of missing data, the score operators are simply the conditional expectations of the score functions in the model for the complete data given the data actually observed. Since tangent sets and tangent spaces are defined in terms of parametric models, it will suffice to work, for the moment at least, in a parametric setting.

Suppose that $X^0 \sim Q_\theta \in \mathcal{Q} \equiv \{Q_\theta : \theta \in \Theta\}$ where $\mathcal{Q}$ is a parametric model. Suppose we observe $X \equiv T(X^0)$ where T is a measurable map from $(\mathcal{X}^0, \mathcal{A}^0)$ to $(\mathcal{X}, \mathcal{A})$. Let the induced model be denoted by $\mathcal{P} \equiv \mathcal{P}_T \equiv \{Q_\theta T^{-1} : \theta \in \Theta\}$. The following basic proposition relates scores for $\mathcal{P} = QT^{-1}$ to the scores for $\mathcal{Q}$.

Proposition 1.1. Suppose that $\mathcal{Q}$ is Hellinger differentiable at $Q \equiv Q_\theta$ for a fixed $\theta \in \Theta$ with score function $a \equiv \dot{\mathbf{l}}(\cdot, Q|\theta, \mathcal{Q})$. Then:

A. $\mathcal{P} = QT^{-1}$ is Hellinger-differentiable at $P = QT^{-1}$ with score function

$$\dot{\mathbf{l}}_\theta(x; \mathcal{P}) \equiv \dot{\mathbf{l}}(x; P|\theta, \mathcal{P}) = E(a(X^0)|T(X^0) = x). \qquad (1.5)$$

B. The information for θ in the model $\mathcal{P} = QT^{-1}$ is always smaller than the information for θ in the model $\mathcal{Q}$:

$$I(\theta; \mathcal{P}) = E\{\dot{\mathbf{l}}_\theta^T(X; \mathcal{P})\dot{\mathbf{l}}_\theta(X; \mathcal{P})\} \leq E\{\dot{\mathbf{l}}_\theta^T(X^0; \mathcal{Q})\dot{\mathbf{l}}_\theta(X^0; \mathcal{Q})\} = I(\theta; \mathcal{Q}).$$

Proof. See e.g. van der Vaart (1988), Bickel, Klaassen, Ritov, and Wellner (1992), or Le Cam and Yang (1988). $\square$

Note that the conditional expectation in (1.5) transforms a function $a = \dot{\mathbf{l}}(\cdot, Q|\theta, \mathcal{Q})$ in $L_2^0(Q)$ to a function $b \equiv \dot{\mathbf{l}}(\cdot, P|\theta, \mathcal{P})$ in $L_2^0(P)$. Thus we can write the score function $\dot{\mathbf{l}}(\cdot, P|\theta, \mathcal{P})$ on the left side of (1.6) as

$$\dot{\mathbf{l}}a \qquad (1.6)$$

where $\dot{\mathbf{l}} \equiv E(\cdot \ |T(X_0))$ is simply the conditional expectation operator. Thus $\dot{\mathbf{l}}$ is a bounded linear map from $L_2^0(Q)$ to $L_2^0(P)$, and it has an *adjoint* (or transpose) $\dot{\mathbf{l}}^T : L_2^0(P) \to L_2^0(Q)$ satisfying

$$< b, \dot{\mathbf{l}}a >_{L_2(P)} = < \dot{\mathbf{l}}^T b, a >_{L_2(Q)}$$

for all $a \in L_2^0(P)$, $b \in L_2^0(Q)$. Then the *information operator* is $\dot{\mathbf{l}}^T \dot{\mathbf{l}}$.

When $\dot{\mathbf{l}}$ is a conditional expectation operator as in (1.5), $\dot{\mathbf{l}}^T$ is also a conditional expectation operator; namely

$$\dot{\mathbf{l}}^T b(x^0) = E(b(X)|X^0 = x^0) - Eb(X).$$

If the model Q is nonparametric or semiparametric, so that $\dot{Q}(Q)$ is $L_2^0(Q)$ or some infinite-dimensional subset thereof,

$$\dot{P}^0 \supset \{\dot{\mathbf{l}}a : a \in \dot{Q}^0\}$$

and

$$\dot{P} \supset \overline{\text{lin}\{\dot{\mathbf{l}}a : a \in \dot{Q}^0\}}.$$

where lin{} denotes the linear span. Usually equality holds, but is more difficult to prove. But from several points of view, it is not necessary to prove the equality.

All of this will become clearer by careful consideration of several examples.

Example 1.1, continued. In this case, $\Theta \subset \mathbb{R}^k$ (a Hilbert space), and we can regard (1.4) as defining a score operator $\dot{\mathbf{l}} : \dot{\Theta} = \mathbb{R}^k \to L_2^0(P_\theta)$ defined by

$$\dot{\mathbf{l}}h = \dot{\mathbf{l}}_\theta h^T \quad for \quad h \in \mathbb{R}^k . \tag{1.7}$$

The adjoint operator $\dot{\mathbf{l}}^T : L_2^0(P_\theta) \to \dot{\Theta} = \mathbb{R}^k$ is defined by

$$\dot{\mathbf{l}}^T b = E_\theta(\dot{\mathbf{l}}_\theta^T b) \quad for \quad b \in L_2^0(P_\theta) , \tag{1.8}$$

and the information operator $\dot{\mathbf{l}}^T \dot{\mathbf{l}}h : \mathbb{R}^k \to \mathbb{R}^k$ is

$$\dot{\mathbf{l}}^T \dot{\mathbf{l}}h = E_\theta(\dot{\mathbf{l}}_\theta^T \dot{\mathbf{l}}_\theta)h^T = I(\theta)h^T. \tag{1.9}$$

Example 1.3, continued. We will regard G as fixed and known. Then we can identify $\dot{Q}$ with $L_2^0(F)$, and, for $a \in L_2^0(F)$, the score operator for this model is given by

$$\begin{aligned}
\dot{\mathbf{l}}a(x) &= E(a(Z)|T(X^0) = x) \\
&= \delta \frac{\int_{[0,u]} a \, dF}{F(u)} + (1-\delta)\frac{\int_{(u,1]} a \, dF}{1 - F(u)}
\end{aligned} \tag{1.10}$$

where $x = (u, \delta)$. By straightforward calculation the adjoint $\dot{\mathbf{l}}^T$ is given by

$$\begin{aligned}
\dot{\mathbf{l}}^T b(z) &= E[b(U,\delta)|Z = z] - Eb(U,\delta) \\
&= \int_0^1 \left\{\{1_{[z \le u]} - F(u)\}b(u,1) + \{1_{[u<z]} - \overline{F}(u)\}b(u,0)\right\} dG(u) \\
&= \int_{[z,1]} b(u,1)dG(u) + \int_{[0,z)} b(u,0)dG(u),
\end{aligned} \tag{1.11}$$

and hence

$$\dot{\mathbf{l}}^T \dot{\mathbf{l}}a(z) = \int_0^1 K(z,t)a(t)dF(t) \tag{1.12}$$

where

$$K(z,t) \equiv \int_0^{t \wedge z} \frac{1}{1 - F(u)}dG(u) + \int_{t \vee z}^1 \frac{1}{F(u)}dG(u) - 1. \tag{1.13}$$

As we will see later, the form (1.12) for the information operator in this model implies that $\dot{\mathbf{l}}^T \dot{\mathbf{l}}$ is *not* boundedly invertible, and that we are, therefore, not allowed to even write down information bounds for an arbitrary (differentiable) function of F. This suggests that many functions of F are *not* estimable at the usual $n^{-1/2}$ rate. We return to a detailed examination of this question in section 3.

Example 1.4, continued. Since G is assumed to be known, we can identify $\dot{Q}$ with $L_2^0(F)$. Then the score operator for this model is given by

$$\dot{\mathrm{l}}a(x) = E[a(Z)|T(X^0) = x] = \frac{\int a(z)g(x-z)dF(z)}{\int g(x-z)dF(z)}. \tag{1.14}$$

By straightforward calculation the adjoint is

$$\dot{\mathrm{l}}^T b(z) = \int b(x)g(x-z)dx - \int b(x)p_F(x)dx$$

$$= \int b(x)\{g(x-z) - p_F(x)\}dx$$

and the information operator $\dot{\mathrm{l}}^T\dot{\mathrm{l}}$ is given by

$$\dot{\mathrm{l}}^T\dot{\mathrm{l}}a(z) = \int K(z,z')a(z')dF(z'). \tag{1.15}$$

where

$$K(z,z') \equiv \int \frac{g(x-z)g(x-z')}{p_F(x)}dx. \tag{1.16}$$

This information operator is of the same basic form as that for example 1.3, with just one integral term: these are smoothing operators with smoothing kernel K. Solving the equation $\dot{\mathrm{l}}^T\dot{\mathrm{l}}a = b$ amounts to solving an integral equation of the *first kind*; see e.g. Tricomi (1957), pages 15 and 143. It will follow in much the same way as for example 1.3 that we cannot give information bounds for arbitrary differentiable functions of F, and this in turn suggests that estimation of many functions of F will not be possible at the rate $n^{-1/2}$.

Example 1.5, continued. In this model, we can again identify $\dot{Q}$ with $L_2^0(F)$. For $a \in L_2^0(F)$ the score operator $\dot{\mathrm{l}}$ is given by

$$\begin{aligned}(\dot{\mathrm{l}}a)(x) &= E[a(Z)|X = x] \\ &= \delta a(t) + (1-\delta)\frac{\int_{(t,\infty)} \frac{1}{z}a(z)dF(z)}{\int_{(t,\infty)} \frac{1}{z}dF(z)}\end{aligned} \tag{1.17}$$

where $x = (t, \delta)$. Then $\dot{\mathrm{l}}$ has adjoint $\dot{\mathrm{l}}^T$ given by

$$(\dot{\mathrm{l}}^T b)(z) = pb(z,1) + (1-p)\frac{1}{z}\int_{[0,z)} b(t,0)dt - Eb(T,\delta),$$

and the information operator is

$$\dot{\mathrm{l}}^T\dot{\mathrm{l}}a(z) = pa(z) + (1-p)\int K(z,z')a(z')dF(z') \tag{1.18}$$

where

$$K(z, z') \equiv \frac{1}{zz'} \int_{[0, z \wedge z')} \frac{1}{\int_{(t,\infty)} u^{-1} dF(u)} dt. \tag{1.19}$$

The information operator for this model is quite different than the information operators which we computed for the models of examples 1.3 and 1.4. It has two terms, so that solving the equation $\dot{\imath}^T \dot{\imath} a = b$ amounts to solving an integral equation of the second kind; see e.g. Tricomi (1957), pages 10, 49, and 63. Operators of this type typically have a bounded inverse. So $(\dot{\imath}^T \dot{\imath})^{-1}$ exists if $p > 0$, and it will turn out that we can write down information bounds for any (differentiable) function ψ of F, suggesting that all such functions can be estimated at rate $n^{-1/2}$ in this model. We will show that all of these claims are indeed true in this particular case.

1.6 Exercises

1. (Convolution with uniform) Let g in example 1.4 be the Uniform$(0,1)$ density $1_{[0,1]}$, and suppose that we know that the mixing distribution function F is also concentrated on $[0,1]$. Then p_F in example 1.4 becomes, with $\delta \equiv 1_{[x \le 1]}$,

$$
\begin{aligned}
p_F(x) &= \int_0^1 1_{[0,1]}(x - z) dF(z) \\
&= \begin{cases} F(x)^\delta (1 - F(x-1))^{1-\delta}, & 0 \le x \le 2 \\ 0, & \text{otherwise} \end{cases}.
\end{aligned}
$$

(a) Show that the score operator for f in this model is given by

$$(\dot{\imath}a)(x) = \frac{\int_0^x a \, dF}{F(x)} 1_{(0,1]}(x) + \frac{\int_{x-1}^1 a \, dF}{1 - F(x-1)} 1_{(1,2]}(x)$$

for $a \in L_2^0(F)$.

(b) Show that the function $x - EX$ is *not* in the range of $\dot{\imath}$ unless F is the Uniform$(0,1)$ distribution.

(c) Show that $\dot{\imath}$ in (a) has adjoint

$$\dot{\imath}^T b(z) = \int_0^1 \{1_{[z \le x]} - F(x)\} b(x) dx + \int_1^2 \{1_{[x-1 \le z]} - (1 - F(x-1))\} b(x) dx$$

and hence that $\dot{\imath}^T$ has nontrivial null space

$$\mathcal{N}(\dot{\imath}^T) = \{b \in L_2(P) : b(x+1) = b(x), 0 \le x \le 1\}.$$

Since $\mathcal{R}(\dot{\imath})^\perp = \mathcal{N}(\dot{\imath}^T)$, it follows that $\overline{\mathcal{R}(\dot{\imath})} \ne L_2^0(P)$.

2. (Properties of adjoints) Let $\mathcal{X}$ and $\mathcal{Y}$ be normed linear spaces with dual spaces $\mathcal{X}^*$ and $\mathcal{Y}^*$ respectively, and suppose that A is a bounded linear operator from $\mathcal{X}$ to $\mathcal{Y}$, $A \in B(\mathcal{X}, \mathcal{Y})$. Recall that

$$||A|| \equiv \sup\{||Ax|| : ||x|| \leq 1\}$$

and that $B(\mathcal{X}, \mathcal{Y})$ is thus a normed space. If $\mathcal{Y}$ is a Banach space, so is $B(\mathcal{X}, \mathcal{Y})$. The *adjoint* or *transpose* A^T of A is defined to be the linear operator from $\mathcal{Y}^*$ to $\mathcal{X}^*$ satisfying

$$< A^T y^*, x >_\mathcal{X} \; = \; < y^*, Ax >_\mathcal{Y}$$

for all $x \in \mathcal{X}$ and $y^* \in \mathcal{Y}^*$. Here $<,>_\mathcal{X}$ on the left side is defined by duality between $\mathcal{X}$ and $\mathcal{X}^*$ so $< A^T y^*, x >_\mathcal{X} \equiv (A^T y^*)(x)$; and $<,>_\mathcal{Y}$ on the right side is defined by duality of $\mathcal{Y}$ and $\mathcal{Y}^*$: $< y^*, Ax >_\mathcal{Y} = y^*(Ax)$. Thus $(A^T y^*)(x) = (y^* A)(x)$ for all $x \in \mathcal{X}$, or $A^T y^* = y^* A$.

Prove the following proposition showing that the transpose operation satisfies the usual properties associated with the transpose of a matrix:

Proposition 1.2.

A. If $I : \mathcal{X} \to \mathcal{X}$ is the identity, $I^T = I : \mathcal{X}^* \to \mathcal{X}^*$.

B. If $A, B \in B(\mathcal{X}, \mathcal{Y})$, then $(A + B)^T = A^T + B^T$.

C. If $A \in B(\mathcal{X}, \mathcal{Y})$ and $c \in R$, then $(cA)^T = cA^T$.

D. If $A \in B(\mathcal{X}, \mathcal{Y})$ and $B \in B(\mathcal{Y}, \mathcal{Z})$, then $(BA)^T = A^T B^T$.

E. If $A \in B(\mathcal{X}, \mathcal{Y})$ has bounded inverse A^{-1}, then $(A^{-1})^T = (A^T)^{-1}$.

F. If $A \in B(\mathcal{X}, \mathcal{Y})$, then $||A^T|| = ||A||$.

3. (The adjoint for an integral type operator)

Suppose that $A : L_2(m_1) \to L_2(m_2)$ is defined by

$$Aa(\cdot) = \int K(s, \cdot) a(s) dm_1(s).$$

Show that if $\int \int |K(s, t)||a(s)b(t)| dm_1(s) dm_2(t) < \infty$ for all $a, b \in L_2(m_1)$, then

$$(A^T b)(\cdot) = \int K(\cdot, t) b(t) dm_2(t).$$

4. Verify the adjoint formula for a conditional expectation operator.

5. Verify the formulas (1.9) and (1.10).

6. Verify the formulas (1.12) and (1.13).

7. Verify the formulas (1.15) and (1.16).

8. Find the score and information operators for example 1.6.

9. Find the score and information operators for example 1.7.

2 Convolution and Asymptotic Minimax Theorems

2.1 Introduction

Now we give statements of several convolution and asymptotic minimax theorems. The key hypotheses involved in virtually all the different formulations of these theorems are:

A. *Local asymptotic normality* (LAN) of the local likelihood ratios of the model. A sufficient condition for this is differentiability of the model in an appropriate sense.

B. For the convolution theorems we will also hypothesize *regularity* of the estimators: the only estimators considered will be those for which the local limiting distributions do not depend on the direction (or magnitude) of the approach of the local parameter point to the fixed parameter point under consideration.

C. *Pathwise differentiability* of the parameter being estimated as a function of the underlying $P \in \mathcal{P}$ with $\mathcal{P}$ metrized by the Hellinger metric. This amounts to Hadamard differentiability along the model $\mathcal{P}$.

Our goal in this chapter will be to explain the basic hypotheses required in different settings, and to discuss several useful refinements and extensions of the basic theorems.

For complete proofs we refer the reader to the primary sources: Hájek (1970, 1972), Le Cam (1972), Ibragimov and Has'minskii (1981), van der Vaart (1988), Bickel, Klaassen, Ritov, and Wellner (1992), van der Vaart and Wellner (1991), Millar (1983), and Le Cam (1986).

2.2 Finite-dimensional parameter spaces

Suppose that $\mathcal{P} = \{P_\theta : \theta \in \Theta\}$, $\Theta \subset I\!\!R^k$ is a Hellinger-differentiable parametric model. With $l(x; \theta) \equiv \log p(x; \theta)$, let

$$L_n(\theta) = \sum_{i=1}^{n} l(X_i; \theta)$$

denote the log-likelihood of $X_1, \ldots, X_n$, a sample from $P_{\theta_0} \equiv P_0 \in \mathcal{P}$. Then it is well-known that

$$L_n(\theta_0 + \frac{t}{\sqrt{n}}) - L_n(\theta_0) = S_n(\theta_0)t^T - \frac{1}{2}tI(\theta_0)t^T + o_{P_0}(1) \tag{2.1}$$

where

$$S_n(\theta_0) \equiv \frac{1}{\sqrt{n}} \sum_{i=1}^{n} \dot{\mathbf{l}}(X_i; \theta_0)$$

is the score for θ at θ_0 (based on the entire sample $X_1, \ldots, X_n$) and $I(\theta_0)$ is the Fisher information matrix defined in (1.3). It follows that

$$L_n(\theta_0 + \frac{t}{\sqrt{n}}) - L_n(\theta_0) \to_d N(-\frac{1}{2}tI(\theta_0)t^T \ , \ tI(\theta_0)t^{T)}) \tag{2.2}$$

under P_0. This is sometimes called the *Local Asymptotic Normality*, or LAN, condition. It is one key ingredient of the Hájek convolution theorem. The second key ingredient is the following definition of regularity of an estimator sequence T_n:

Definition 2.1. $T = \{T_n\}$ is a *locally regular estimator* of θ at $\theta = \theta_0$ if, for every sequence $\{\theta_n\} \subset \Theta$ with $\sqrt{n}(\theta_n - \theta_0) \to t \in \mathbb{R}^k$, under P_{θ_n}

$$\sqrt{n}(T_n - \theta_n) \to_d \mathbb{Z} \text{ as } n \to \infty$$

where the law of $\mathbb{Z}$ depends on θ_0 but not on t. Thus the limit distribution of $\sqrt{n}(T_n - \theta_n)$ does not depend on the direction of approach t of θ_n to θ_0.

With these two basic ingredients, we can state a simplified version of Hájek's (1970) convolution theorem:

Theorem 2.1. (Hájek). Suppose that (2.2) holds with $I(\theta_0)$ nonsingular and that $\{T_n\}$ is a regular estimator of θ at θ_0. Then

$$\mathbb{Z} \stackrel{d}{=} \mathbb{Z}_0 + \Delta_0 \tag{2.3}$$

where $\mathbb{Z}_0 \sim N(0, I^{-1}(\theta_0))$ is independent of Δ_0.

Hájek (1970) proved a somewhat more general theorem based on just the LAN hypothesis (2.2) using a method based on "Bayesian considerations." A different proof using characteristic functions due to Peter Bickel may be found in Roussas (1972), and also in Bickel, Klaassen, Ritov, and Wellner (1992). This latter type of proof was exploited and developed by R. Beran (1977a, 1977b) in more general settings.

In words, theorem 2.1 says that the limiting distribution of any regular estimator T_n of θ must be at least as "spread out" as the $N(0, I^{-1}(\theta_0))$ distribution of $\mathbb{Z}_0$. Thus an *efficient estimator* is a regular estimator for which the limiting distribution is exactly equal to $\mathbb{Z}_0$. Another way to say this is in terms of the following asymptotic optimality theorem:

Corollary 2.1. (Hájek, 1970). Suppose that $\{T_n\}$ is a locally regular estimator of θ at θ_0 and that $l : \mathbb{R}^k \to \mathbb{R}^+$ is bowl-shaped: i.e.

(i) $l(x) = l(-x)$.

(ii) $\{x : l(x) \le c\}$ is convex for every $c \ge 0$.

Then

$$\liminf_{n\to\infty} E_{\theta_0} l(\sqrt{n}(T_n - \theta_0)) \geq El(\mathbb{Z}_0) \tag{2.4}$$

where $\mathbb{Z}_0 \sim N(0, I^{-1}(\theta_0))$.

If a supremum over θ in a local neighborhood of θ_0 is added to the left side of (2.4), then the same type of statement holds for an *arbitrary* (not necessarily regular) estimator T_n of θ. This is the Hájek-Le Cam asymptotic minimax theorem due to Hájek (1971), and, in a more abstract form, to Le Cam (1971).

Theorem 2.2. (Hájek, 1971). Suppose that (2.2) holds, that T_n is any estimator of θ, and that l is bowl-shaped. Then

$$\lim_{b\to\infty} \liminf_{n\to\infty} \sup_{\{\theta: \sqrt{n}|\theta-\theta_0|\leq b\}} E_\theta l(\sqrt{n}(T_n - \theta)) \geq El(\mathbb{Z}_0) \tag{2.5}$$

More generally, suppose that $\{P_\theta^n : \theta \in \Theta, n = 1, 2, \ldots\}$ is a family of distributions indexed by $\Theta \subset \mathbb{R}^k$ (not necessarily the distributions of independent observations) satisfying

$$\frac{dP_{\theta_0+u\psi(n)}^n}{dP_{\theta_0}^n} = \exp\{S_n u^T - \frac{1}{2}|u|^2 + R_n(\theta_0, u)\} \tag{2.6}$$

where

$$S_n \to_d Z \sim N_k(0, J) \text{ under } P_{\theta_0}^n, \tag{2.7}$$

J denotes the $k \times k$ identity matrix, $\psi(n) = \psi(n, \theta_0)$ is a $k \times k$ matrix, and

$$R_n(\theta_0, u) \to_p 0 \text{ under } P_{\theta_0}^n. \tag{2.8}$$

Then, the asymptotic minimax theorem holds in the following strengthened form:

Theorem 2.3. (Ibragimov and Has'minskii, 1981). Suppose that LAN in the form (2.6)-(2.8) holds and that l is bowl-shaped. Then for any estimator T_n of θ,

$$\liminf_{n\to\infty} \sup_{\{\theta: \psi^{-1}(n)(\theta-\theta_0)\in K_b\}} E_\theta l(\psi^{-1}(n)(T_n - \theta))$$

$$\geq \begin{cases} (1 - b^{-1/2})^k El(Z)1_{[Z\in K_{\sqrt{b}}]} \\ \\ \frac{1}{2^k} El(Z)1_{[Z\in K_{b/2}]} \end{cases} \tag{2.9}$$

where $Z \sim N_k(0, J)$ and K_b is the cube in $\mathbb{R}^k$ with vertices having all coordinates $\pm b$.

Proof. See Ibragimov and Has'minskii (1981), equations (12.18) and (12.19), page 168 and 169. $\square$

2.3 Infinite-dimensional parameter spaces

Theorems 2.1 and 2.2 can be generalized greatly in several different directions. In particular, if we replace $\Theta \subset \mathbb{R}^k$ by a linear subspace $\mathcal{H}$ of some Hilbert space and replace estimation of $\theta \in \Theta$ by estimation of $\nu_n(P) \in \mathcal{B}$, a Banach space, then these theorems have very natural extensions within the context of local asymptotic normality (i.e. Gaussian limit experiments). The theorems we give here, based on work of van der Vaart and Wellner (1991), extend earlier results of Koshevnik and Levit (1976), Levit (1978), and Millar (1983). For other extensions, to non-Gaussian limit experiments, we refer the interested reader to Le Cam (1971), Jeganathan (1981, 1982), Millar (1984), Strasser (1985), and van der Vaart (1990).

We will write $<,>$ for the inner product in $\mathcal{H}$, and $||\cdot||_{\mathcal{H}}$ for the norm in $\mathcal{H}$. We will also write $||\cdot||$ for the norm in $\mathcal{B}$. As usual we need hypotheses guaranteeing:

  A. LAN or differentiability of the model;
  B. Regularity of the estimators considered;
  C. Differentiability of the parameter(s) $\nu = \nu(P)$ being considered.

Here are the particular forms of these hypotheses which we will use here:

  A. (LAN). For $n = 1, 2, \ldots$ and $h \in \mathcal{H}$, let $P_{n,h}$ be a probability measure defined on a measurable space $(\mathcal{X}_n, \mathcal{A}_n)$. Assume that

$$\log \frac{dP_{n,h}}{dP_{n,0}} = S_{n,h} - \frac{1}{2}||h||_{\mathcal{H}}^2 + o_{P_{n,0}}(1) \qquad (2.10)$$

  for every $h \in \mathcal{H}$ where $S_{n,h} : \mathcal{X}_n \to R$ are measurable maps with

$$(S_{n,h_1}, \ldots, S_{n,h_d}) \to_d N_d(0, <h_i, h_j>) \text{ under } P_{n,0} \qquad (2.11)$$

  for every finite subset $h_1, \ldots, h_d \in \mathcal{H}$.

  B. (Regularity of estimators). A sequence of maps $T_n : \mathcal{X}_n \to \mathcal{B}$ is said to be *regular for* ν_n if, under $P_{n,h}$

$$R_n(T_n - \nu_n(P_{n,h})) \Rightarrow \mathbb{Z}, \text{ as } n \to \infty$$

  for every $h \in \mathcal{H}$ where $\mathbb{Z}$ is a Borel measurable, tight random element in $\mathcal{B}$ which does not depend on $h \in \mathcal{H}$ and $R_n : \mathcal{B} \to \mathcal{B}$ is a sequence of linear maps with $||R_n|| \to \infty$. Here $\Rightarrow$ is in the sense of Hoffmann-Jørgensen (1984); see van der Vaart and Wellner (1991) for a review of this theory.

  C. (Differentiability of $\nu_n = \nu_n(P_{n,h})$). The parameter $\nu_n(P_{n,h}) \in \mathcal{B}$ satisfies

$$R_n(\nu_n(P_{n,h}) - \nu_n(P_{n,0})) \to \dot{\nu}(h) \text{ as } n \to \infty \qquad (2.12)$$

  for every $h \in \mathcal{H}$ where $\dot{\nu} : \mathcal{H} \to \mathcal{B}$ is a continuous linear map.

Here is a convolution theorem based on these hypotheses from van der Vaart and Wellner (1991).

Theorem 2.4. (van der Vaart and Wellner, 1991). Suppose (2.10)–(2.12) hold and that $\{T_n\}$ is regular. Then there exist tight Borel measurable elements $\mathbb{Z}_0$ and Δ_0 in $\mathcal{B}$ with:

A. $\mathbb{Z} =^d \mathbb{Z}_0 + \Delta_0$.

B. $\mathbb{Z}_0$ and Δ_0 are independent.

C. $b^*\mathbb{Z}_0 \sim N(0, \|\dot{\nu}^T b^*\|^2_{\mathcal{H}})$ for every $b^* \in \mathcal{B}^*$.

Proof. See van der Vaart and Wellner (1991), section 2. $\square$

Much as in section 2.2, there is a related asymptotic optimality theorem for regular estimators; see e.g. Bickel, Klaassen, Ritov, and Wellner (1992), section 5.2.

To state an analogue of theorem 2.2, we need to add a hypothesis guaranteeing some asymptotic measurability of our estimators T_n as follows: Given a linear subspace $\mathcal{B}'$ of $\mathcal{B}^*$, let $T_n : \mathcal{X}_n \to \mathcal{B}$ be maps (estimators) such that

$$E_0^* f(R_n(T_n - \nu_n(P_{n,0}))) - E_{0,*} f(R_n(T_n - \nu_n(P_{n,0}))) \to 0 \qquad (2.13)$$

for every f of the form $f(x) = g(b'_1 x, \ldots, b'_m x)$ with $b'_1, \ldots, b'_m \in \mathcal{B}'$ and $g \in C_b(I\!\!R^m)$.

Assume that there exists a tight Borel measurable random element $\mathbb{Z}_0$ in $\mathcal{B}$ such that

$$b^*\mathbb{Z}_0 \sim N(0, \|\dot{\nu}^T b^*\|^2_{\mathcal{H}}) \text{ for every } b^* \in \mathcal{B}^*. \qquad (2.14)$$

We call a function $l : \mathcal{B} \to R$ subconvex with respect to a topology τ on $\mathcal{B}$ if

$$\begin{array}{c} l(0) = 0 \leq l(x) \text{ for every} x \in \mathcal{B} \\ l(x) = l(-x) \text{ for every } x \in \mathcal{B} \\ \{x : l(x) \leq c\} \text{ is convex and } \tau - \text{closed for every } c \in R. \end{array} \qquad (2.15)$$

We let $\tau(\mathcal{B}')$ be the weakest topology on $\mathcal{B}$ that makes $b' : \mathcal{B} \to R$ continuous for all $b' \in \mathcal{B}'$. Here is a very natural generalization (and extension) of theorems 2.2 and 2.3:

Theorem 2.5. (van der Vaart and Wellner, 1991). Suppose that (2.10)–(2.14) hold. Then for every $\tau(\mathcal{B}')$-subconvex function l, and an arbitrary sequence of estimators $\{T_n\}$,

$$\sup_{I \subset \mathcal{H}} \liminf_{n \to \infty} \sup_{h \in I} E_{h,*} l(R_n(T_n - \nu_n(P_{n,h}))) \geq El(\mathbb{Z}_0) \qquad (2.16)$$

where the first supremum is over all finite subsets I of $\mathcal{H}$.

Proof. See van der Vaart and Wellner (1991). $\square$

It will be helpful to recast theorem 2.4 in the special case of iid observations, and to connect it with the tangent space considerations of section 1. Let $\mathcal{P}$ be a collection of probability measures on a measurable space $(\mathcal{X}, \mathcal{A})$, and let $X_1, \ldots, X_n$ be iid $P \in \mathcal{P}$. Then the tangent space $\dot{\mathcal{P}}(P) = \dot{\mathcal{P}} \subset L_2^0(P)$ at P as defined in lecture 1.4 will play the role of the Hilbert space $\mathcal{H}$ in theorem 2.4. Now we write $\| \cdot \|_2$ for the $L_2(P)$ metric on $\dot{\mathcal{P}}$. Then the Hellinger differentiable submodels $\mathcal{P}_0$ of $\mathcal{P}$ yield sequences $\{P_n\} \equiv \{P_{\theta_n}\}$ in $\mathcal{P}$ with $\theta_n = \theta + tn^{-1/2}$ satisfying

$$\int \{\sqrt{n}[(dP_n)^{1/2} - (dP)^{1/2}] - \frac{1}{2}h(dP)^{1/2}\}^2 \to 0 \qquad (2.17)$$

for $h = \dot{l}t^T$ in the tangent set $\dot{\mathcal{P}}^0$. We say that $\nu : \mathcal{P} \to \mathcal{B}$ is *pathwise differentiable* at $P \in \mathcal{P}$ if there is a continuous linear map $\dot{\nu}$ from $\dot{\mathcal{P}}^0$ to $\mathcal{B}$ such that

$$\sqrt{n}(\nu(P_n) - \nu(P)) \to \dot{\nu}(h) \qquad (2.18)$$

for every sequence $\{P_n\} \subset \mathcal{P}$ satisfying (2.17). Further, an estimator sequence $\{T_n\}$ of $\nu(P)$ is *regular at P* if, under P_n^n for every sequence $\{P_n\}$ satisfying (2.17),

$$\sqrt{n}(T_n - \nu(P_n)) \Rightarrow \mathbb{Z} \text{ as } n \to \infty \qquad (2.19)$$

where $\mathbb{Z}$ does not depend on h. With these specializations of our earlier definitions, we obtain the following corollary of theorem 2.4:

Corollary 2.2. Suppose that:

 (i) ν is pathwise differentiable at $P \in \mathcal{P}$ with derivative $\dot{\nu}$.
 (ii) $\{T_n\}$ is regular with limit $\mathbb{Z}$.
(iii) $\dot{\mathcal{P}}^0$ is linear.

Then there exist tight Borel measurable random elements $\mathbb{Z}_0$ and Δ_0 in $\mathcal{B}$ such that:

 A. $\mathbb{Z} =^d \mathbb{Z}_0 + \Delta_0$.
 B. $\mathbb{Z}_0$ and Δ_0 are independent.
 C. $b^*\mathbb{Z}_0 \sim N(0, \|\dot{\nu}^T b^*\|_2^2)$ for every $b^* \in \mathcal{B}^*$ where $\dot{\nu}^T : \mathcal{B}^* \to \dot{\mathcal{P}}$.

Proof. This follows immediately from theorem 2.4 with the identification $\mathcal{H} = \dot{\mathcal{P}}$. $\square$

It is important and useful to connect corollary 2.2 with the parametric forms of the convolution theorems in section 2.1. The key connecting link is immediate once we recognize that calculation of $\dot{\nu}^T$ — which is crucially involved in the description of the optimal variance appearing in part C — essentially involves projection onto $\dot{\mathcal{P}}$. Suppose that ν_e is an extension of ν to $\mathcal{M} \equiv \{\text{all } P \text{ on } (\mathcal{X}, \mathcal{A})\}$. If ν_e is pathwise differentiable at $P \in \mathcal{M}$, we obtain an extension $\dot{\nu}_e$ of $\dot{\nu}$: i.e. $\dot{\nu}_e : L_2^0(P) \to \mathcal{B}$ and $\dot{\nu}_e|_{\dot{\mathcal{P}}} = \dot{\nu}$. Then we claim that

$$\dot{\nu}^T b^* = \Pi(\dot{\nu}_e^T b^* | \dot{\mathcal{P}}) \text{ for all } b^* \in \mathcal{B}^*. \qquad (2.20)$$

This follows since, for $h \in \dot{\mathcal{P}}$, $b^* \in \mathcal{B}^*$,

$$
\begin{aligned}
< \dot{\nu}^T b^*, h > &= < \Pi(\dot{\nu}_e^T b^* | \dot{\mathcal{P}}), h > \quad \text{by (2.20)} \\
&= < \dot{\nu}_e^T b^* - (\dot{\nu}_e^T b^* - \Pi(\dot{\nu}_e^T b^* | \dot{\mathcal{P}})), h > \\
&= < \dot{\nu}_e^T b^*, h > \quad \text{since} \quad \dot{\nu}_e^T b^* - \Pi(\dot{\nu}_e^T b^* | \dot{\mathcal{P}}) \perp \dot{\mathcal{P}} \\
&= < b^*, \dot{\nu}_e(h) > \quad \text{since} \quad \dot{\nu}_e^T \text{ is the adjoint of } \dot{\nu}_e \\
&= < b^*, \dot{\nu}(h) > \quad \text{since} \quad \dot{\nu}_e|_{\dot{\mathcal{P}}} = \dot{\nu} .
\end{aligned}
$$

Thus the way we calculate an optimal variance of an estimator of ν for a given model $\mathcal{P}$ can be viewed as follows: we first calculate the derivative $\dot{\nu}_e$ of the function considered as a map on the bigger space $\mathcal{M}$. Then we project $\dot{\nu}_e^T b^*$ onto the tangent space $\dot{\mathcal{P}}$ of the model of interest. The squared length of this projection yields the variance of an efficient estimator of $\nu(P)$.

In view of this connection, we introduce the notation $\tilde{\mathbf{l}}_\nu$ for $\dot{\nu}^T$, and call $\tilde{\mathbf{l}}_\nu \equiv \dot{\nu}^T : \mathcal{B}^* \to \dot{\mathcal{P}}$ the *efficient influence operator* for estimation of ν. (In the classical parametric case with $\nu \equiv \theta$, $\mathcal{B} = \mathbb{R}^k = \mathcal{B}^*$, $\tilde{\mathbf{l}}_\nu = \dot{\mathbf{l}} I^{-1}(\nu)$ so $\tilde{\mathbf{l}}_\nu b^* = \dot{\mathbf{l}} I^{-1}(\nu) b^{*T}$.)

Example 2.1. (Example 1.1, continued; estimation of the mean of the Weibull distribution). This continues examples 1.2.1 and 1.3.1. Suppose that $\mathcal{P}$ is the Weibull family given in example 1.2.1; we calculated the scores and information matrix in example 1.3.1. Consider estimation of $\nu(P_\theta) = E_\theta X = \alpha \Gamma(\frac{1}{\beta} + 1)$ where $\Gamma(x) \equiv \int_0^\infty t^{x-1} e^{-t} dt$ is the usual gamma function. Now $\nu_e(P) \equiv E_P X$ is pathwise differentiable on $\mathcal{M}_\lambda$ $ca \{ P : E_P X^2 \leq M \}$ with derivative $\dot{\nu}_e : L_2^0(P) \to R$ given by

$$
\begin{aligned}
\dot{\nu}_e(h) = E_P \dot{\nu}_e^T(X) h(X) &= \int_0^\infty (x - EX) h(x) dP(x) \\
&= < \dot{\nu}_e^T , h >_P .
\end{aligned}
$$

Note that $\dot{\nu}_e^T(x) = x - EX \notin \dot{\mathcal{P}} = [\dot{\mathbf{l}}_\alpha, \dot{\mathbf{l}}_\beta]$ if $\beta \neq 1$. Thus

$$
\begin{aligned}
(\dot{\nu}^T 1)(x) &= \Pi(\dot{\nu}_e^T 1 | \dot{\mathcal{P}}) \\
&= \Pi(\dot{\nu}_e^T 1 | [\dot{\mathbf{l}}_\alpha, \dot{\mathbf{l}}_\beta]) \\
&= Cov_\theta(X - E_\theta X, \dot{\mathbf{l}}_\theta) I_\theta^{-1} \dot{\mathbf{l}}_\theta(x) \\
&= \frac{6}{\pi^2} \frac{\alpha}{\beta} \{ c(t - 1) - d(1 - logt\{t - 1\}) \}.
\end{aligned}
$$

where $t \equiv (x/\alpha)^\beta$, and $c(\beta) \equiv c \equiv b\Gamma(\frac{1}{\beta} + 1) + a\Gamma'(\frac{1}{\beta} + 1)$, $d(\beta) \equiv d \equiv a\Gamma(\frac{1}{\beta} + 1) + \Gamma'(\frac{1}{\beta} + 1)$.

Furthermore the information lower bound for estimation of ν is

$$
I_\nu^{-1} \equiv \|\dot{\nu}^T 1\|_2^2 = Cov_\theta(X - E_\theta X, \dot{\mathbf{l}}_\theta) I_\theta^{-1} Cov_\theta(X - EX, \dot{\mathbf{l}}_\theta)^T
$$

$$\begin{aligned} = \ &\frac{6}{\pi^2}\frac{\alpha^2}{\beta^2}\left\{b\Gamma^2(\frac{1}{\beta}+1)+2a\Gamma\Gamma'(\frac{1}{\beta}+1)+(\Gamma'(\frac{1}{\beta}+1))^2\right\} \\ < \ &\|\dot{\nu}_e^T 1\|_2^2 = E_\theta(X-E_\theta X)^2 \\ = \ &Var_\theta(X) = \alpha^2\{\Gamma(\frac{2}{\beta}+1)-\Gamma^2(\frac{1}{\beta}+1)\}. \end{aligned}$$

Example 2.2. (Estimation of the mean of a normal distribution). Suppose that $\mathcal{P} = \{N(\theta, 1) : \theta \in \Theta = R\}$. Then $\dot{l}_\theta(x) = x - \theta$, $I(\theta) = 1$, and $\dot{\mathcal{P}} = [\dot{l}_\theta]$ is a one dimensional subspace of $L_2^0(P_\theta)$. One extension of $\nu(P_\theta) = \theta$ for $P_\theta \in \mathcal{P}$ is $\nu_e(P) = F_P^{-1}(1/2)$ where F_P is the distribution function associated with P on R $(F_P(x) = P(X \le x))$, and F_P^{-1} denotes the left-continuous inverse of F_P. Then ν_e is pathwise differentiable on $\mathcal{M}_\lambda^+$, the collection of distributions P on R with positive, continuous density p at $F_P^{-1}(1/2)$, with derivative $\dot{\nu}_e : L_2^0(P) \to R$ given by

$$\dot{\nu}_e(h) = E_P \dot{\nu}_e^T(X)h(X) \quad = \quad < \dot{\nu}_e^T , h >_P$$

where

$$\dot{\nu}_e^T(x) = \frac{-1}{p(F_P^{-1}(1/2))}(1_{(-\infty, \nu_e(P)]}(x) - \frac{1}{2}).$$

Note that $\dot{\nu}_e^T \notin \dot{\mathcal{P}}$. Then, for $P = P_\theta \in \mathcal{P}$, $\dot{\nu}^T = \Pi(\dot{\nu}_e^T | \dot{\mathcal{P}})$ is given by

$$\dot{\nu}^T(x) = \frac{<\dot{\nu}_e^T, \dot{l}_\theta >_P}{\|\dot{l}_\theta\|_2^2}\dot{l}_\theta(x) = x - \theta,$$

and

$$\begin{aligned} I_\nu^{-1} \equiv \|\dot{\nu}^T 1\|_2^2 \ &= \ E_\theta(X-\theta)^2 \\ &= \ 1 \\ &< \ \|\dot{\nu}_e^T 1\|_2^2 \\ &= \ \frac{E_\theta(1_{(-\infty, \nu_e(P)]}(x) - \frac{1}{2})^2}{p^2(F_P^{-1}(1/2))} \\ &= \ \frac{\pi}{2}. \end{aligned}$$

2.4 Exercises

The following sequence of exercises is aimed at producing an asymptotic minimax bound for estimation of the distribution function F at a single point in example 1.3.

1. (Interval censoring). Consider the conditional distribution of δ given U in example 1.2.3. It follows easily that the conditional distribution of δ given U is *Bernoulli(F(U))*.

Consider the following parametric submodels of $\mathcal{P}$: Fix $t_0 \in (0,1)$ and $c > 0$, and let

$$J_n \equiv [t_0 - cn^{-1/3}, t_0 + cn^{-1/3}],$$
$$h_n(u) \equiv f(t_0)\{1_{[t_0 - cn^{-1/3}, t_0]}(u) - 1_{(t_0, t_0 + cn^{-1/3}]}(u)\},$$
$$H_n(t) \equiv \int_0^t 1_{J_n}(u) h_n(u) du.$$

Finally, for $|\theta| \leq M < 1$, let

$$F_n(t; \theta) \equiv F(t) + \theta H_n(t). \tag{2.21}$$

Suppose that $(X_1, \ldots, X_n) \equiv ((U_1, \delta_1), \ldots, (U_n, \delta_n))$ are iid $P_{F_n, G}$ with $F_n \equiv F_n(\cdot, \theta)$ as in (2.21). Show that the Fisher information for θ from $(\delta_i | U_i)$ is $H_n^2(U_i)/(F_n(U_i; \theta)(1 - F_n(U_i; \theta)))$ for $i = 1, \ldots, n$, and hence the information for θ from $(\delta_1, \ldots, \delta_n | U_1, \ldots, U_n)$ is

$$I_n(\theta) \equiv \sum_{i=1}^n I_{ni}(\theta) = \sum_{i=1}^n \frac{H_n^2(U_i)}{F_n(U_i; \theta)(1 - F_n(U_i; \theta))}.$$

2. (Interval censoring, continued). Show that for almost all sequences $U_1, U_2, \ldots$ the conditional information $I_n(\theta)$ of problem 1 satisfies

$$I_n(\theta) \equiv \frac{1}{\psi^2(n)} \rightarrow \frac{2}{3} c^3 f^2(t_0) g(t_0) \frac{1}{F(t_0)(1 - F(t_0))}.$$

3. (Interval censoring, continued). Let P_θ^n in (2.6)–(2.8) denote the conditional distribution of $\delta_1, \ldots, \delta_n$ given $U_1, \ldots, U_n$ under $P_{F_n, G} \equiv P_{F_n(\cdot, \theta), G}$. Show that with $\psi^2(n) = 1/I_n(\theta)$ as in problem 3, (2.6)–(2.8) hold (with $k = 1$).

4. (Interval censoring, continued). Apply the second form of the strengthened local asymptotic minimax theorem 2.3 to conclude that, for any estimator U_n of $F(t_0)$,

$$\sup_{\delta \in (0,1)} \liminf_{n \to \infty} \sup_{\theta : |\theta| \leq \delta} E_\theta |n^{1/3}(U_n - F_n(t_0; \theta)|$$

$$\geq \frac{1}{\sqrt{2\pi}} \frac{12^{1/3}}{8^{1/2}} (.79133 \ldots)[F \bar{F} \frac{f}{g}(t_0)]^{1/3}.$$

where $M \equiv .79133 \ldots$ is the maximum value of $a^{-1/2}(1 - e^{-a^3})$, achieved by the solution of $e^{a^3} - 6a^2 - 1 = 0$, or $a = a_0 \equiv 1.429 \ldots$.

5. Repeat the calculations of example 2.1 with $\nu(P_\theta) = P_\theta(X \geq t_0) = e^{-(\frac{t_0}{\alpha})^\beta}$ for a fixed number t_0. [Hint: Use the fact that $\nu_e(P) \equiv P(X \geq t_0)$ has $\dot{\nu}_e$ given by $\dot{\nu}_e(h) = E_P(h(1_{[t_0, \infty)} - P_\theta(X \geq t_0)) = < h, \dot{\nu}_e^T >_P$; and hence $\dot{\nu}_e^T(x) = 1_{[t_0, \infty)}(x) - P_\theta(X \geq t_0).]$

3 Van der Vaart's Differentiability Theorem

3.1 Differentiability of Implicitly Defined Functions

Frequently the model $\mathcal{P}$ is described directly in terms of natural finite or infinite-dimensional parameters; recall example 1.1 and examples 1.3–1.6. Then we want to estimate certain functionals of the underlying parameter. For example, in example 1.1.3 we may want to estimate the mean of F, or F at a single point t_0. Thus, in terms of the distribution $P = P_{F,G}$ of the observed data, we want to estimate the implicitly defined functional

$$\nu(P_{F,G}) = \psi(F) \tag{3.1}$$

where ψ is the mean of F or $F(t_0)$.

The key question we want to address in this lecture is: when is an implicitly defined functional ν of this type (3.1) pathwise differentiable in the sense of (2.18)? When pathwise differentiability of ν holds, then we can use corollary 2.2 to give a convolution — information lower bound — for estimation of ν, suggesting that ν will be estimable at rate $n^{-1/2}$. When differentiability of ν fails, then the convolution theorem does not apply, and this implies in turn that we should only expect estimation of ν with some slower rate of convergence, and often with a non-Gaussian limit distribution.

To make this precise, suppose that the model $\mathcal{P}$ is parametrized by a subset $\mathcal{G}$ of a Hilbert space $\mathcal{H}$: $\mathcal{P} = \{P_g : g \in \mathcal{G}\}$ where $\mathcal{G} \subset \mathcal{H}$. Fix $g \in \mathcal{G}$, and suppose that $\mathcal{G}_g$ is the collection of all curves or paths $\{g_\eta\} \subset \mathcal{G}$ through g with

$$g_\eta = g + \eta h + o(\eta) \text{ in } \mathcal{H} \tag{3.2}$$

for some $h \in \mathcal{H}$. Then let

$$\dot{\mathcal{G}}^0 = \cup_{\{g_\eta\} \in \mathcal{G}_g} \{h : (3.2) \text{ holds for } \{g_\eta\}\}.$$

For simplicity we suppose that $\mathcal{P}$ is dominated by some fixed measure μ, and let $p_g \equiv dP_g/d\mu$, $s(g) \equiv s_g \equiv \sqrt{p_g}$. As before, we suppose that $\mathcal{B}$ is a fixed Banach space.

Here are our assumptions about $\dot{\mathcal{G}}^0$ and differentiability of the model as a function of g:

A. $\dot{\mathcal{G}}^0$ is a closed and linear subspace of $\mathcal{H}$. Thus $\dot{\mathcal{G}} \equiv [\dot{\mathcal{G}}^0] = \dot{\mathcal{G}}^0$.
B. There is a bounded linear operator $\dot{\mathbf{l}} : \dot{\mathcal{G}}^0 \to L_2^0(P_g)$ such that for every $\{g_\eta\}$ in $\mathcal{G}_g$ satisfying (3.2),

$$s(g_\eta) - s(g) - \eta(\frac{1}{2}\dot{\mathbf{l}}h)s(g) = o(\eta) \text{ in } L_2(\mu). \tag{3.3}$$

When (3.3) holds, we say that $\mathcal{P}$ is *Hellinger differentiable* at P_g along $\{P_{g_\eta}\}$, and we call $\dot{\mathbf{l}}$ the *score operator* for g.

C. The parameter $\nu : \mathcal{P} \to \mathcal{B}$ is given by

$$\nu(P_g) = \psi(g) \tag{3.4}$$

where $\psi : \mathcal{G} \to \mathcal{B}$ is pathwise differentiable at g with derivative $\dot{\psi}_g \equiv \dot{\psi} : \dot{\mathcal{G}}^0 \to \mathcal{B}$.

The following theorem is due to van der Vaart (1991).

Theorem 3.1. (van der Vaart). Suppose that A–C hold.

(i) If $\nu : \mathcal{P} \to \mathcal{B}$ given by (3.4) is pathwise differentiable at $P_g \in \mathcal{P}$, then

$$\mathcal{R}(\dot{\psi}^T) \subset \mathcal{R}(\dot{\mathbf{i}}^T). \tag{3.5}$$

(ii) If (3.5) holds and $\dot{\mathcal{P}} = \overline{\mathcal{R}(\dot{\mathbf{i}})}$, then ν is pathwise differentiable and the efficient influence operator $\tilde{\mathbf{i}}_\nu$ for estimation of ν is the unique operator from $\mathcal{B}^*$ to $\dot{\mathcal{P}}$ satisfying

$$\dot{\psi}^T = \dot{\mathbf{i}}^T \tilde{\mathbf{i}}_\nu. \tag{3.6}$$

The following diagram may help in understanding the various operators involved in theorem 1.

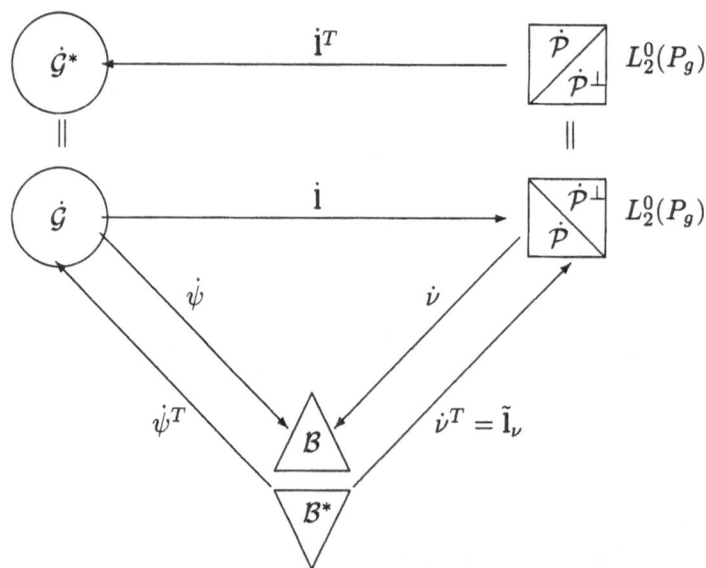

Remark 1.1. For any bounded linear operator A from a Hilbert space $\mathcal{X}$ to another Hilbert space $\mathcal{Y}$, $\mathcal{N}(A^T) = \mathcal{R}(A)^\perp$ and $\mathcal{N}(A) = \mathcal{R}(A^T)^\perp$. Thus $\mathcal{R}(\dot{\psi}^T)^\perp = \mathcal{N}(\dot{\psi})$ and $\mathcal{R}(\dot{\mathbf{i}}^T)^\perp = \mathcal{N}(\dot{\mathbf{i}})$ by e.g. Rudin (1973), theorem 4.12, page 94, so (3.5) implies

$$\mathcal{N}(\dot{\mathbf{i}}) \subset \mathcal{N}(\dot{\psi}). \tag{3.7}$$

Necessity of (3.7) is intuitively clear by reasoning as follows. Suppose that (3.7) fails: then there is an $h \in \mathcal{N}(\dot{1}) \setminus \mathcal{N}(\dot{\psi})$ with $\psi(g_\eta) = \psi(g) + \eta\dot{\psi}(h) + o(\eta)$, but by (3.3) and $\dot{1}h = 0$, $P_{g_\eta} = P_g + o(\eta)$ in Hellinger distance. Thus P_{g_η} is very close to P_g while $\psi(g_\eta)$ remains distant from $\psi(g)$, and hence we cannot tell the difference between $\psi(g_\eta)$ and $\psi(g)$ based on observations from P_{g_η}.

Proof. Suppose that ν is pathwise norm-differentiable. Then by (2.18), pathwise differentiability of ψ, and assumption B,

$$\dot{\nu}(\dot{1}h) = \lim_{\eta \to 0} \frac{\nu(P_{g_\eta}) - \nu(P_g)}{\eta} = \dot{\psi}(h) \qquad (a)$$

for every $h \in \dot{\mathcal{G}}^0$. This extends by continuity to all $h \in \dot{\mathcal{G}}$. Therefore we have continuous linear maps $\dot{\psi} : \dot{\mathcal{G}} \to \mathcal{B}$, $\dot{1} : \dot{\mathcal{G}} \to \dot{\mathcal{P}}$, and $\dot{\nu} : \dot{\mathcal{P}} \to \mathcal{B}$, such that

$$\dot{\psi} = \dot{\nu}\dot{1} . \qquad (b)$$

Hence

$$\dot{\psi}^T = \dot{1}^T \dot{\nu}^T \qquad (c)$$

by exercise 1.2 part D, and (3.5) follows.

Conversely, if (3.5) holds, define $\dot{\nu} : \dot{\mathcal{P}}^0 \to \mathcal{B}$ on $\dot{\mathcal{P}}^0 \equiv \mathcal{R}(\dot{1})$ by

$$\dot{\nu}(\dot{1}h) = \dot{\psi}(h) \text{ for } h \in \dot{\mathcal{G}}. \qquad (d)$$

This is well-defined, since, if $\dot{1}h_1 = \dot{1}h_2$, then $h_1 - h_2 \in \mathcal{N}(\dot{1}) \subset \mathcal{N}(\dot{\psi})$ by (3.7), and hence $\dot{\psi}(h_1) = \dot{\psi}(h_2)$. Furthermore $\dot{\nu}$ is a linear map on $\mathcal{R}(\dot{1}) = \dot{\mathcal{P}}^0$.

It remains only to show that $\dot{\nu}$ is continuous (i.e. bounded). For any $b^* \in \mathcal{B}^*$ it follows from (d) that

$$\begin{aligned} b^*\dot{\nu}(\dot{1}h) &= <b^*, \dot{\psi}h>_{\mathcal{B}} = <\dot{\psi}^T b^*, h> \\ &= <\dot{1}^T a, h> \text{ by (3.5) for some } a = a(b^*) \in L_2(P_g) \\ &= <a, \dot{1}h>_0 . \end{aligned}$$

Thus $b^*\dot{\nu}$ is a continuous linear (real-valued) functional on $\mathcal{R}(\dot{1})$ for every $b^* \in \mathcal{B}^*$. Boundedness of $\dot{\nu} : \mathcal{R}(\dot{1}) = \dot{\mathcal{P}}^0 \to \mathcal{B}$ follows from an application of the Banach-Steinhaus theorem; see exercise 3.2 (with $\mathcal{X} \equiv \mathcal{R}(\dot{1})$ and $\mathcal{Y} \equiv \mathcal{B}$). Now extend $\dot{\nu}$ to $\dot{\mathcal{P}} = \overline{\mathcal{R}(\dot{1})}$ by continuity.

Finally, (3.6) is implied by (e) and the definition of $\tilde{1}_\nu$: $\tilde{1}_\nu b^* = \Pi(\dot{\nu}^T(P_0)b^* | \dot{\mathcal{P}})$.

It remains only to prove uniqueness. Suppose that $\tilde{1}_\nu^{(1)}, \tilde{1}_\nu^{(2)}$ are both solutions with $d \equiv \tilde{1}_\nu^{(1)}(b^*) - \tilde{1}_\nu^{(2)}(b^*) \neq 0$ for some $b^* \in \mathcal{B}^*$. Note that $d \in \dot{\mathcal{P}}$. Then $\dot{1}^T(d) = 0$, and

$$0 = <\dot{1}^T(d), h> = <d, \dot{1}h>_0$$

for all $h \in \dot{\mathcal{G}}$. Thus $d \perp \dot{\mathcal{P}}$, a contradiction. We conclude that $\tilde{1}_\nu$ solving (3.6) is unique. $\square$

Corollaries and consequences

The key hypothesis (3.5) of theorem 1 can be strengthened in a variety of ways to yield sufficient conditions. The following corollary was established by van der Vaart (1988a).

Corollary 3.1. Suppose that A–C hold. Then $\nu : \mathcal{P} \to \mathcal{B}$ is pathwise norm-differentiable at $P_g \in \mathcal{P}$ if

$$\mathcal{N}(\dot{1}) \subset \mathcal{N}(\dot{\psi}) \tag{3.8}$$

and

$$\mathcal{R}(\dot{1}) \ \text{ is closed} . \tag{3.9}$$

Proof. By theorem 3.1 it suffices to show that (3.8) and (3.9) imply (3.5). By definition of $\dot{\psi}^T b^*$ (recall (b) of the proof of theorem 1), we have $\dot{\psi}^T b^* \perp \mathcal{N}(\dot{\psi})$. Thus by (3.8) and remark 1

$$\dot{\psi}^T b^* \in \mathcal{N}(\dot{\psi})^\perp \subset \mathcal{N}(\dot{1})^\perp = \overline{\mathcal{R}(\dot{1}^T)} = \mathcal{R}(\dot{1}^T)$$

where the last equality follows since $\mathcal{R}(\dot{1})$ closed if and only if $\mathcal{R}(\dot{1}^T)$ is closed; see e.g. Rudin (1973), theorem 4.14, page 96. □ Hence (3.5) holds. □

However, (3.9) fails quite frequently as we will see in the examples in the next section. For a simple example in which (3.9) fails, see the indicator censoring example in the next section. Another class of examples for which (3.9) typically fails are the convolution models, example 1.4. The Vardi-Zhang model, example 1.5, provides an example in which (3.9) holds.

Corollary 3.1 replaces (3.5) by its implication (3.7) or (3.8) and closedness of the range of $\dot{1}$; these conditions are clearly only sufficient since (3.9) does not involve the particular function ψ. The following corollary 3.2 goes further, and replaces both (3.8) and (3.9) by conditions independent of the particular function ψ. It is essentially in the spirit of the hypotheses of section 4 of Begun, Hall, Huang, and Wellner (1983).

Corollary 3.2. Suppose that A–C hold. Then:

(i) $\nu : \mathcal{P} \to \mathcal{B}$ is pathwise norm-differentiable at $P_g \in \mathcal{P}$ if

$$\mathcal{N}(\dot{1}) = 0 \tag{3.10}$$

and

$$\mathcal{R}(\dot{1}) \ \text{ is closed.} \tag{3.11}$$

(ii) $\dot{1}^T \dot{1}$ is one-to-one and onto if and only if (3.10) and (3.11) hold.

(iii) $\mathcal{R}(\dot{1}^T \dot{1}) \subset \mathcal{R}(\dot{1}^T)$ with equality if and only if (3.11) holds.

When (3.9) fails, we can work instead with the condition (3.5) and "solve" (3.6): i.e. given $b^* \in \mathcal{B}^*$, find the solution $\tilde{l}_\nu b^* \in \dot{\mathcal{P}} \subset L_2(P_g)$ of

$$\dot{\psi}^T b^* = \dot{l}^T(\tilde{l}_\nu b^*). \tag{3.12}$$

Alternatively, in view of corollary 2.iii, we can replace (3.5) by the sufficient condition

$$\mathcal{R}(\dot{\psi}^T) \subset \mathcal{R}(\dot{l}^T \dot{l}) \tag{3.13}$$

and find h^* to solve, for fixed $b^* \in \mathcal{B}^*$,

$$\dot{\psi}^T b^* = \dot{l}^T \dot{l} h^* \tag{3.14}$$

for $h^* \in \dot{\mathcal{G}}$. Note that (3.14) is similar to the usual "normal equations" of linear regression theory, $X^T X \beta = X^T Y$ with $\dot{l}$ playing the role of X, h^* replacing β, and $\dot{\psi}^T b^*$ replacing $X^T Y$.

Write h^* satisfying (3.14) as

$$h^* = (\dot{l}^T \dot{l})^- \dot{\psi}^T b^*.$$

Then

$$\tilde{l}_\nu b^* = \dot{l}(\dot{l}^T \dot{l})^- \dot{\psi}^T b^* \in \mathcal{R}(\dot{l}) \subset \dot{\mathcal{P}} . \tag{3.15}$$

Proof of corollary 3.2. (i) follows immediately from corollary 3.1 since (3.10) implies (3.8) trivially.

To prove (ii), note that if $\dot{l}^T \dot{l}$ is one-to-one, then so is $\dot{l}$. If $\dot{l}^T \dot{l}$ is onto, then $\dot{l}^T$ is onto and $\mathcal{R}(\dot{l}^T) = \dot{\mathcal{G}}$ is closed by A. Thus $\mathcal{R}(\dot{l})$ is closed too, as in the proof of corollary 3.1. Conversely, if $\dot{l}^T \dot{l} h = 0$, then $\dot{l}h \in \mathcal{N}(\dot{l}^T) = \mathcal{R}(\dot{l})^\perp$ by remark 3.1. Since $\dot{l}h \in \mathcal{R}(\dot{l})$, this implies $\dot{l}h = 0$, and hence $h = 0$ if $\mathcal{N}(\dot{l}) = \{0\}$. That $\dot{l}^T \dot{l}$ is onto is a consequence of (iii) since (3.11) then implies $\mathcal{R}(\dot{l}^T \dot{l}) = \mathcal{R}(\dot{l}^T)$, where

$$\mathcal{R}(\dot{l}^T) = \overline{\mathcal{R}(\dot{l}^T)} = \mathcal{N}(\dot{l})^\perp = \{0\}^\perp.$$

The inclusion in (iii) is obvious. If $\mathcal{R}(\dot{l})$ is closed, suppose $h = \dot{l}^T \alpha \in \mathcal{R}(\dot{l}^T)$. Then

$$\begin{aligned}
h = \dot{l}^T \alpha &= \dot{l}^T(\Pi(\alpha|\mathcal{R}(\dot{l})) + \Pi(\alpha|\mathcal{R}(\dot{l})^\perp)) \\
&= \dot{l}^T(\dot{l}h_1 + \Pi(\alpha|\mathcal{N}(\dot{l}^T))) \text{ by remark 1} \\
&= \dot{l}^T \dot{l} h_1 \text{ for some } h_1 \in \dot{\mathcal{G}} ,
\end{aligned}$$

so $h \in \mathcal{R}(\dot{l}^T \dot{l})$.

Conversely, let $\alpha \in \overline{\mathcal{R}(\dot{l})}$. Consider $\dot{l}^T \alpha$; by $\mathcal{R}(\dot{l}^T \dot{l}) = \mathcal{R}(\dot{l}^T)$ there exists h_0 such that $\dot{l}^T \alpha = \dot{l}^T \dot{l} h_0$. Hence $\alpha - \dot{l} h_0 \in \mathcal{N}(\dot{l}^T) = \mathcal{R}(\dot{l})^\perp$. On the other hand $\alpha - \dot{l} h_0 \in \overline{\mathcal{R}(\dot{l})}$. Thus $\alpha - \dot{l} h_0 = 0$, or $\alpha \in \mathcal{R}(\dot{l})$; hence $\mathcal{R}(\dot{l})$ is closed. $\square$

The following proposition gives relationships between various ranges and closures of ranges of the score and information operators. It will be useful in connection with convolution models in the next section.

Proposition 3.1. Suppose that $\mathcal{P}$ is any model with score operator $\dot{\imath}$. Then:

(i) $\mathcal{R}(\dot{\imath})$ is dense in $L_2^0(P)$ if and only if $\dot{\imath}^T$ is one-to-one (i.e. $\mathcal{N}(\dot{\imath}^T) = \{0\}$).

(ii) If $\dot{\imath}^T$ is one-to-one, then $\overline{\mathcal{R}(\dot{\imath})} = L_2^0(P) = \dot{\mathcal{M}}$.

(iii) If $\dot{\imath}$ is one-to-one, then $\mathcal{R}(\dot{\imath}^T\dot{\imath}) = \dot{\mathcal{G}}$ if and only if $(\dot{\imath}^T\dot{\imath})^{-1}$ is bounded.

Proof. (i) follows from remark 3.1: $\mathcal{N}(\dot{\imath}^T) = \mathcal{R}(\dot{\imath})^\perp$; see e.g. Rudin (1973), page 94. Part (ii) is an immediate consequence of (i).

To prove part (iii), first note that $\mathcal{N}(\dot{\imath}^T\dot{\imath}) = \mathcal{N}(\dot{\imath})$: $\mathcal{N}(\dot{\imath}) \subset \mathcal{N}(\dot{\imath}^T\dot{\imath})$ is obvious; if $h \in \mathcal{N}(\dot{\imath}^T\dot{\imath})$, then

$$0 = <\dot{\imath}^T\dot{\imath}h, h >_G = <\dot{\imath}h, \dot{\imath}h >_0 = ||\dot{\imath}h||_0^2$$

so that $\dot{\imath}h = 0$ a.e. P_0, and hence $h \in \mathcal{N}(\dot{\imath})$. Since $\dot{\imath}$ is one-to-one by hypothesis, $\dot{\imath}^T\dot{\imath}$ is also one-to-one. Now by remark 3.1 again, the self-adjoint operator $\dot{\imath}^T\dot{\imath}$ has range which is dense in $L_2^0(G)$. By Banach's theorem (see exercise 3.3), $(\dot{\imath}^T\dot{\imath})^{-1}$ is bounded if and only if the range of $\dot{\imath}^T\dot{\imath}$ is equal to $L_2^0(G)$. $\square$

3.2 Some Applications of the Differentiability Theorem

Now we apply theorem 3.1 to several of the examples introduced in section 1.2.

Example 3.1. (Interval Censoring, continued). This is a continuation of example 1.3.

The density p_F of (U, δ) is

$$p_F(x) = p_F(u, \delta) = F(u)^\delta (1 - F(u))^{1-\delta} g(u).$$

Recall that we are considering G as fixed and known. This model is Hellinger differentiable with respect to $f^{1/2}$ in the sense of (3.3) with score operator $\dot{\imath}$ given by (1.10):

$$\dot{\imath}a(x) = E(a(Z)|T(X^0) = x)$$
$$= \delta \frac{\int_{[0,u]} a \, dF}{F(u)} + (1-\delta)\frac{\int_{(u,1]} a \, dF}{1-F(u)} \tag{3.16}$$

where $x = (u, \delta)$. Then $\dot{\imath}$ has adjoint $\dot{\imath}^T$ given by

$$\dot{\imath}^T b(z) = E[b(U,\delta)|Z = z] - Eb(U,\delta)$$
$$= \int_0^1 \left\{ \{1_{[z \le u]} - F(u)\}b(u,1) + \{1_{[u<z]} - \bar{F}(u)\}b(u,0)\right\} dG(u) \tag{3.17}$$
$$= \int_{[z,1]} b(u,1)dG(u) + \int_{[0,z)} b(u,0)dG(u),$$

and information operator

$$\mathbf{i}^T \mathbf{i} a(z) = \int_0^1 K(z,t)a(t)dF(t) \tag{3.18}$$

where K is as given in (1.10). It follows that $\int K(z,t)a(t)dF(t) = 0$ and by straightforward calculation that

$$\int_0^1 \int_0^1 K^2(s,t)dF(s)dF(t) = 2\int_0^1 \int_0^v \frac{F(u)(1-F(v))}{F(v)(1-F(u))}dG(u)dG(v) \le 1. \tag{3.19}$$

Thus $\mathbf{i}^T \mathbf{i}$ is a Hilbert-Schmidt operator (see e.g. Reed and Simon (1972), theorem VI.23, page 210). In particular $\mathbf{i}^T \mathbf{i}$ is compact and $(\mathbf{i}^T \mathbf{i})^{-1}$ does not exist as a bounded operator. Therefore corollary 3.2 does *not* apply.

Consider estimation of the distribution function F at a fixed point t_0: i.e.

$$\nu(P_F) \equiv \psi(f^{1/2}) \equiv \int_0^{t_0}(f^{1/2})^2 d\mu = F(t_0).$$

Then $\psi : \dot{\mathcal{G}} = L_2^0(F) \to \mathcal{B} = R$ is given by

$$\dot{\psi}a = \int_0^1 [1_{[0,t_0]}(z) - F(t_0)]a(z)dF(z)$$

for $a \in \dot{\mathcal{G}}$, and

$$(\dot{\psi}^T c)(z) = c(1_{[0,t_0]}(z) - F(t_0)), \quad c \in R.$$

This is a discontinuous function of z for every c, but all the functions $\mathbf{i}^T b$ are continuous functions of z since G is continuous. Therefore (3.5) fails for this ψ, and hence, by theorem 3.1, $\nu(P_F) = F(t_0)$ is *not pathwise differentiable* at any $P \in \mathcal{P}$. Thus the convolution theorem, corollary 2.2 does not apply, and we conclude that $F(t_0)$ cannot be estimated at rate $n^{-1/2}$.

But now consider estimation of the fixed linear functional

$$\psi(f^{1/2}) = \int a(z)(f^{1/2}(z))^2 dz = \int a(z)dF(z),$$

where a is given. Then

$$(\dot{\psi}^T)c(z) = c(a(z) - \int adF), \quad c \in R,$$

and the condition (3.5) for differentiability becomes, in view of (3.17),

$$a(z) - \int adF = E(b(U,\delta)|Z=z)$$

for some $b \in L_2^0(P_F)$. In particular, if $a(z) = z$ so that ψ is the mean of F, then

$$\dot{\psi}a = \int_0^1 (z - EZ)a(z)dF(z),$$

$$(\dot{\psi}^T 1)(z) = z - EZ,$$

and it is easy to see that $\dot{\psi}^T 1$ is in the range of $\dot{l}^T$ if

$$\int_0^{1/2} \frac{F(z)}{g(z)}dz + \int_{1/2}^1 \frac{\bar{F}(z)}{g(z)}dz < \infty.$$

In particular,

$$\tilde{l}_\nu(u, \delta) = \begin{cases} -\bar{F}(u)/g(u), & \delta = 1 \\ F(u)/g(u), & \delta = 0 \end{cases}$$

satisfies $\dot{l}^T \tilde{l}_\nu = \dot{\psi}^T$ and $\tilde{l}_\nu \in \overline{\mathcal{R}(\dot{l})}$. (In fact, $\tilde{l}_\nu \in \mathcal{R}(\dot{l})$; note that $a \equiv [-(F\bar{F}/g)']/f$ works!) Thus

$$I_\nu^{-1} \equiv E(\tilde{l}_\nu^2) = \int_0^1 \frac{F(u)(1 - F(u))}{g(u)}du.$$

As shown in part II, section 5.4, this information bound is achieved by the mean of the nonparametric maximum likelihood estimator $\hat{F}_n$ of F.

Example 3.2. (Gaussian convolution model). This is a continuation and special case of example 1.3. The density of X is

$$p_F(x) = \int g(x - z)dF(z)$$

where g is the standard Gaussian density ϕ. This model is Hellinger differentiable with respect to $f^{1/2}$ in the sense of (3.3) with score operator given by (1.11):

$$\dot{l}a(x) = E[a(Z)|T(X^0) = x] = \frac{\int a(z)g(x - z)dF(z)}{\int g(x - z)dF(z)}. \tag{3.20}$$

As calculated in section 1.5, the adjoint $\dot{l}^T$ of $\dot{l}$ is

$$\dot{l}^T b(z) = \int b(x)g(x - z)dx - \int b(x)p_F(x)dx \tag{3.21}$$

$$= \int b(x)\{g(x - z) - p_F(x)\}dx$$

for $b \in L_2^0(P)$, and the information operator $\dot{l}^T \dot{l}$ is given by

$$\dot{l}^T \dot{l}a(z) = \int K(z, z')a(z')dF(z'). \tag{3.22}$$

where K is given in (1.13).

Properties of these operators depend considerably on the particular fixed density g and on the support of F. If g is the Uniform$(0,1)$ density, then by exercise 1.1, $\mathcal{N}(\dot{1}^T) \neq 0$, and hence $\overline{\mathcal{R}(\dot{1})} \neq L_2^0(P)$. Frequently, however $\dot{1}^T$ is one-to-one so that $\mathcal{N}(\dot{1}^T) = \{0\}$ and $\overline{\mathcal{R}(\dot{1})} = L_2^0(P)$ by proposition 3.1.ii. But if $g = \phi$, the Gaussian case, and the support of F has non-empty interior, then by standard exponential family theory both $\mathcal{N}(\dot{1}) = \{0\}$ and $\mathcal{N}(\dot{1}^T) = \{0\}$, so that $\overline{\mathcal{R}(\dot{1}^T)} = L_2^0(F)$ and $\overline{\mathcal{R}(\dot{1})} = L_2^0(P_F)$.

Consider estimation of the distribution function F at a fixed point t_0: i.e.

$$\nu(P_F) \equiv \psi(f^{1/2}) \equiv \int_0^{t_0} (f^{1/2})^2 d\mu = F(t_0).$$

Just as in example 3.1, $\psi : \dot{\mathcal{G}} = L_2^0(F) \to \mathcal{B} = R$ is given by

$$\dot{\psi}a = \int_0^1 [1_{[0,t_0]}(z) - F(t_0)]a(z)dF(z)$$

for $a \in \dot{\mathcal{G}}$, and

$$(\dot{\psi}^T c)(z) = c(1_{[0,t_0]}(z) - F(t_0)), c \in R.$$

This is a discontinuous function of z for every c. But it is clear from the form of $\dot{1}^T$ that all the functions in $\mathcal{R}(\dot{1}^T)$ are continuous functions, and hence $\dot{\psi}^T$ is not in the range of $\dot{1}^T$. It follows that (3.5) fails for this ψ, and hence $\nu(P_F) = F(t_0)$ is *not pathwise differentiable* at any $P \in \mathcal{P}$. Thus the convolution theorem, corollary 2.3.1 does not apply and we conclude that $F(t_0)$ is not regularly estimable at rate $n^{-1/2}$.

As in example 3.1, consider estimation of a fixed linear functional $\psi(f^{1/2}) = \int a(z)dF(z)$ where a is given. Then $\dot{\psi}^T 1(z) = a(z) - \int a dF$ and the condition (3.5) for differentiability becomes, in view of (3.21),

$$a(z) - \int a dF = \int b(x)g(x-z)dx - \int b(x)p_F(x)dx$$

for some $b \in L_2(P_F)$, or $a(z) = \int b(x)g(x-z)dx$. If this holds, then a regular estimator of ψ is given by $n^{-1} \sum_{i=1}^n b(X_i)$.

3.3 Exercises

1. (Adjoints and orthgonal complements) Show that if $\mathcal{X}$ and $\mathcal{Y}$ are Hilbert spaces, and $A \in B(\mathcal{X}, \mathcal{Y})$, then

$$\mathcal{N}(A^T) = \mathcal{R}(A)^\perp \text{ and } \mathcal{N}(A) = \mathcal{R}(A^T)^\perp$$

where $\perp$ denotes orthogonal complement.

2. (A consequence or variant of the Banach-Steinhaus theorem) Suppose that $\mathcal{X}$ is a normed linear space and $\mathcal{Y}$ is a Banach space. If $f : \mathcal{X} \to \mathcal{Y}$ satisfies $y^* \circ f \in \mathcal{X}^*$ for every $y^* \in \mathcal{Y}^*$, then f is continuous and linear. (Reference: see sections 2.4 and 2.6 in Rudin (1973).)

3. (Banach's theorem) Prove the following theorem: If $\mathcal{X}$ and $\mathcal{Y}$ are Banach spaces and the linear map $A : \mathcal{X} \to \mathcal{Y}$ is one-to-one ($\mathcal{N}(A) = \{0\}$), then A^{-1} is continuous (i.e. bounded) if and only if $\mathcal{R}(A)$ is closed. (Reference: Jörgens (1982), theorem 5.1, page 81; or, Rudin (1973), corollary 2.12, page 28.)

4. Verify equation (3.19).

Part II

Nonparametric

Maximum Likelihood Estimation

1 The Interval Censoring Problem

1.1 Characterization of the non-parametric maximum likelihoo destimators

We study the following two cases of interval censoring.

Case 1. Let $(X_1, T_1), \ldots, (X_n, T_n)$ be a sample of random variables in $\mathbb{R}_+^2$, where X_i and T_i are independent (non-negative) random variables with distribution functions F_0 and G, respectively. The only observations which are available are T_i ("observation time") and $\delta_i = \{X_i \leq T_i\}$. Here and (often) in the sequel we will denote the indicator of an event A (such as $\{X_i \leq T_i\}$) just by A, instead of 1_A. The log likelihood for F_0 is given by the function

$$F \mapsto \sum_{i=1}^{n} \left\{ \delta_i \log F(T_i) + (1 - \delta_i) \log\big(1 - F(T_i)\big) \right\}, \tag{1.1}$$

where F is a right-continuous distribution function.

Case 2. Let $(X_1, T_1, U_1), \ldots, (X_n, T_n, U_n)$ be a sample of random variables in $\mathbb{R}_+^3$, where X_i is a (non-negative) random variable with distribution function F_0, and where T_i and U_i are (non-negative) random variables, independent of X_i, with a joint distribution function H and such that $T_i \leq U_i$ with probability one. The only observations which are available are (T_i, U_i) (the "observation times") and $\delta_i = \{X_i \leq T_i\}$, $\gamma_i = \{X_i \in (T_i, U_i]\}$. In this case the log likelihood for F_0 is given by the function

$$F \mapsto \sum_{i=1}^{n} \left\{ \delta_i \log F(T_i) + \gamma_i \log\big(F(U_i) - F(T_i)\big) + (1 - \delta_i - \gamma_i) \log\big(1 - F(U_i)\big) \right\}. \tag{1.2}$$

The second case of interval censoring is related to "double censoring", which is studied e.g., in Chang (1990), but the important difference is that with interval censoring the value of X_i is unknown, even if we know that $T_i < X_i \leq U_i$, whereas in the case of double censoring the value of X_i *is* known if $T_i < X_i \leq U_i$. The remaining parts of the models for double censoring and for interval censoring, case 2, are the same.

We first study the likelihood equations for interval censoring, case 1. The log likelihood, divided by n, can be written in the following way:

$$\psi(F) \overset{\text{def}}{=} \int_{\mathbb{R}^2} \left\{ 1_{\{x \leq t\}} \log F(t) + 1_{\{x > t\}} \log\{1 - F(t)\} \right\} dP_n(x, t), \tag{1.3}$$

where P_n is the empirical probability measure of the pairs (X_i, T_i), $1 \leq i \leq n$. The *nonparametric maximum likelihood estimator* (NPMLE) $\hat{F}_n$ of F is a (right-continuous) distribution function F, maximizing (1.3).

Remark 1.1. Note that only the values of $\hat{F}_n$ at the observation points matter for the maximization problem. To avoid trivialities, we will take as "the" NPMLE a distribution function which is piecewise constant, and only has jumps at the observation points. It may happen that the likelihood function is maximized by a function F such that $F(t) < 1$, at each observation point t. In this case we do not specify the location of the remaining mass to the right of the biggest observation point. We shall show that, under these conventions, the NPMLE is uniquely determined, both in case 1 and case 2 of the interval censoring problem.

An interesting example of interval censoring, Case 1, is discussed by Niels Keiding (1991), who analyzes data, obtained by K. Dietz. The data consist of a cross-sectional survey where people of various ages had their blood tested for antibodies against the Hepatitis A virus. The NPMLE of the (sub)distribution function of the ages at which people are infected by the virus is shown in Figure 1.

step function: NPMLE
dashed curve: integrated density estimate

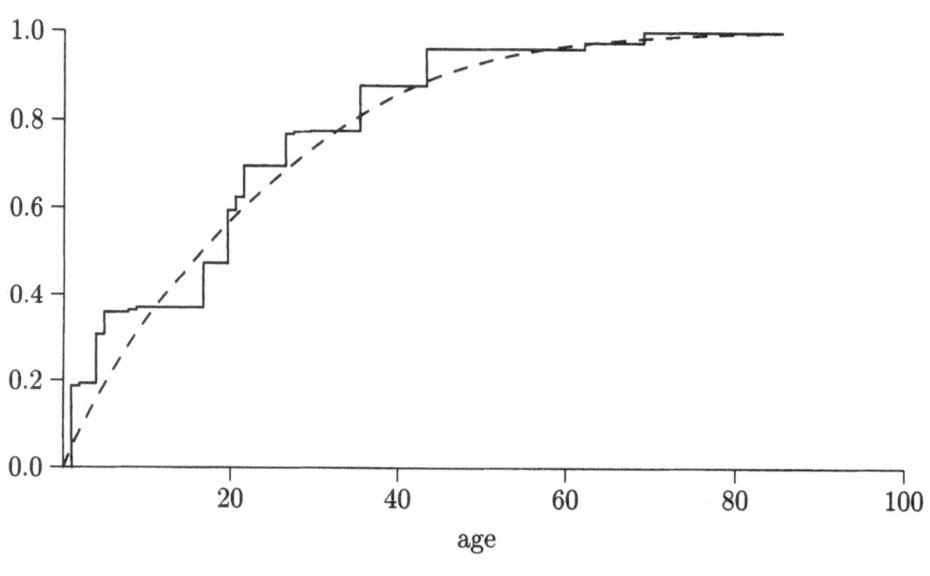

age

Figure 1

This is a clear example of interval censoring, Case 1, since the antibodies only show that the tested person has been infected by the virus at some unknown time in the past, whereas the absence of antibodies shows that he/she did not get the disease and may still get it (or may never get it). The estimates of the density and the hazard rate, taken from Groeneboom (1991), are shown in Figures 2 and 3. They are obtained by estimating the density by

$$\hat{f}_n(x) = h(x)^{-1} \int K_x((a-t)/h(x)) \, d\hat{F}_n(t),$$

where K_x is a so-called "boundary kernel". The hazard rate is then estimated by

$$\hat{\lambda}_n(x) = \hat{f}_n(x)/\{1 - \tilde{F}_n(x)\},$$

where $\tilde{F}_n$ is the integrated density estimate (also shown in Figure 1). For more details on this, we refer to Groeneboom (1991) and Keiding (1991).

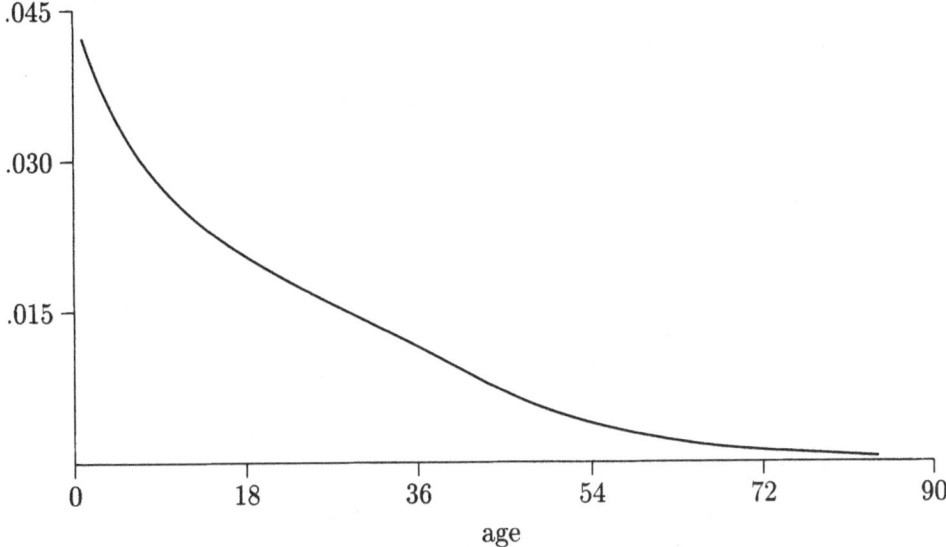

Figure 2. Density estimate $\hat{f}_n$

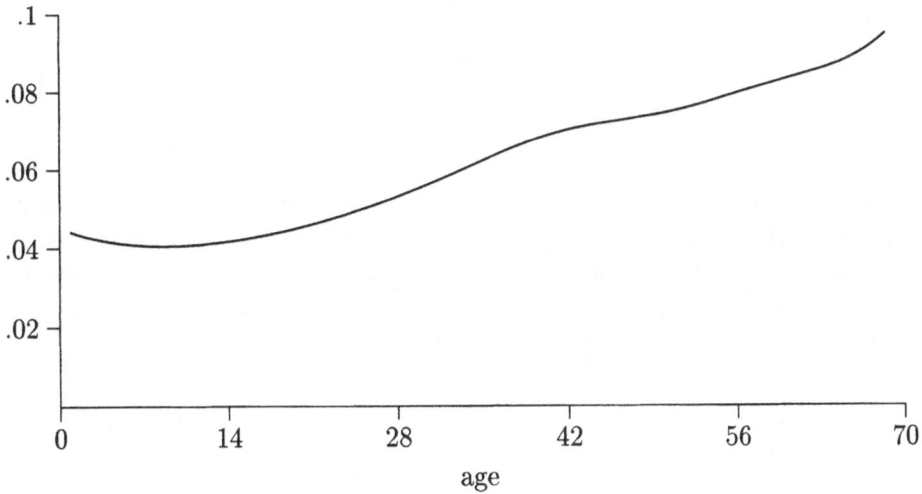

Fig. 3. Estimate of the hazard rate

We now first give the usual characterization of the NPMLE in terms of the so-called self-consistency equations. This characterization is used by Turnbull (1974) who attributes the concept to Efron (1967). Let, for each $t \in \mathbb{R}$, the function $A_t : \mathbb{R} \to \mathbb{R}$ be defined by

$$A_t(u) = \begin{cases} 0, & \text{if } F(t) = 0, \ u \in \mathbb{R}, \\ F(u \wedge t)/F(t) - F(u), & \text{if } F(t) > 0, \ u \in \mathbb{R}, \end{cases} \qquad (1.4)$$

where $u \wedge t = \min\{u, t\}$. Differentiating the function ψ in the direction A_t yields:

$$\lim_{h \downarrow 0} \frac{\psi(F + hA_t) - \psi(F)}{h}$$

$$= \int_{\mathbb{R}^2} \left\{ 1_{\{x \le u\}} \frac{F(u \wedge t)/F(t) - F(u)}{F(u)} \right.$$

$$\left. - 1_{\{x > u\}} \frac{F(u \wedge t)/F(t) - F(u)}{1 - F(u)} \right\} \, dP_n(x, u). \qquad (1.5)$$

If F is the NPMLE, then (1.5) should equal zero, for each t. So we get for the NPMLE $\hat{F}_n$ the equation:

$$\hat{F}_n(t) = \int_{\mathbb{R}^2} \left\{ 1_{\{x \le u\}} \frac{\hat{F}_n(u \wedge t)}{\hat{F}_n(u)} + 1_{\{x > u\}} \frac{\hat{F}_n(t) - \hat{F}_n(u \wedge t)}{1 - \hat{F}_n(u)} \right\} \, dP_n(x, u).$$

A nicer way of writing this equation is:

$$\hat{F}_n(t) = E_{\hat{F}_n} \left\{ F_n(t) \mid T_1, \ldots, T_n, \delta_1, \ldots, \delta_n \right\}, \qquad (1.6)$$

where F_n is the (unobservable) empirical distribution function of the random variables $X_1, \ldots, X_n$. So $\hat{F}_n(t)$ is the conditional expectation of the empirical distribution function F_n at t, given the available information $T_1, \ldots, T_n, \delta_1, \ldots, \delta_n$, under the (self-induced) probability measure $P_{\hat{F}_n}$. We will see in Chapter 3 that the fixed point equation (1.6) immediately yields the iteration steps of the EM algorithm. We note in passing here that the "self-consistency equation" (1.6) does not uniquely determine the NPMLE $\hat{F}_n$, even under the conventions of Remark 1.1.

We now want to give a different characterization of the NPMLE, using concepts from the theory of isotonic regression. For this, some notation is needed. Let $T_{(i)}$ be the ith order statistic of $T_1, \ldots, T_n$, and let $\delta_{(i)}$ be the corresponding indicator, i.e., if $T_j = T_{(i)}$, then $\delta_{(i)} = 1_{\{X_j \le T_j\}}$. The NPMLE corresponds to a vector $\tilde{y} = (y_1, \ldots, y_n) \in \mathbb{R}^n$, maximizing the function

$$\phi(\tilde{x}) = \sum_{i=1}^{n} \left\{ \delta_{(i)} \log x_i + (1 - \delta_{(i)}) \log(1 - x_i) \right\}, \quad \tilde{x} \in \mathbb{R}^n, \qquad (1.7)$$

under the side condition

$$0 \le x_1 \le \ldots \le x_n \le 1. \qquad (1.8)$$

Note that if $\delta_{(i)} = 0$, $i = 1, \ldots, k$, the $\tilde{y}$, maximizing (1.7) should satisfy $y_1 = \ldots = y_k = 0$, since this makes the corresponding second term in the sum (1.7) as big as possible and puts no additional constraints on the values of y_i, for $i > k$. Likewise, if $\delta_{(i)} = 1$, $j \leq i \leq n$, the $\tilde{y}$, maximizing (1.7) should satisfy $y_j = \ldots = y_n = 1$.

So in the maximization problem we may assume without loss of generality that $\delta_{(1)} = 1$ and $\delta_{(n)} = 0$. We then can also assume without loss of generality that $y_1 > 0$ and $y_n < 1$, if $\tilde{y}$ maximizes (1.7), since otherwise we would have $\phi(\tilde{y}) = -\infty$.

The following proposition gives necessary and sufficient conditions for $\tilde{y}$ to be a vector, maximizing (1.7), under the constraint (1.8) and the just mentioned restrictions.

Proposition 1.1. Let $\delta_{(1)} = 1$ and $\delta_{(n)} = 0$, and let $\tilde{y} = (y_1, \ldots, y_n)$ satisfy (1.8), with x_i replaced by y_i. Then $\tilde{y}$ maximizes (1.7) if and only if

$$\sum_{j \geq i} \left\{ \frac{\delta_{(j)}}{y_j} - \frac{1 - \delta_{(j)}}{1 - y_j} \right\} \leq 0, \; i = 1, \ldots, n, \tag{1.9}$$

and

$$\sum_{i=1}^{n} \left\{ \frac{\delta_{(i)}}{y_i} - \frac{1 - \delta_{(i)}}{1 - y_i} \right\} y_i = 0. \tag{1.10}$$

Moreover, $\tilde{y}$ is uniquely determined by (1.9) and (1.10).

Proof. First suppose that $\tilde{y}$ satisfies (1.9) and (1.10). Since we may assume $y_1 > 0$ and $y_n < 1$, all terms in (1.9) are finite. The function ϕ is concave, so if $\tilde{x}$ satisfies (1.8), we get

$$\phi(\tilde{x}) - \phi(\tilde{y}) \leq \langle \nabla \phi(\tilde{y}), \tilde{x} - \tilde{y} \rangle, \tag{1.11}$$

where $\nabla \phi(\tilde{y})$ is the vector of partial derivatives

$$\nabla \phi(\tilde{y}) = \left(\frac{\delta_{(1)}}{y_1} - \frac{1 - \delta_{(1)}}{1 - y_1}, \ldots, \frac{\delta_{(n)}}{y_n} - \frac{1 - \delta_{(n)}}{1 - y_n} \right).$$

Furthermore, $\langle \nabla \phi(\tilde{y}), \tilde{y} \rangle = 0$, if $\tilde{y}$ satisfies (1.10).

Now note that each $\tilde{x}$, satisfying (1.8), can be written in the following form

$$\tilde{x} = \sum_{i=1}^{n} \alpha_i \tilde{1}_i, \tag{1.12}$$

where $\alpha_i = x_{n-i+1} - x_{n-i}$, $\tilde{1}_i$ is a vector which has 1's as its last i components and zeros as its first $n - i$ components, and where $x_0 \stackrel{\text{def}}{=} 0$. Hence, if $\tilde{x}$ satisfies

(1.8), we get

$$\langle \nabla \phi(\tilde{y}), \tilde{x} - \tilde{y} \rangle = \langle \nabla \phi(\tilde{y}), \tilde{x} \rangle = \sum_{i=1}^{n} \alpha_i \langle \nabla \phi(\tilde{y}), \tilde{1}_i \rangle$$

$$= \sum_{i=1}^{n} \alpha_i \sum_{j \geq i} \left\{ \frac{\delta_{(j)}}{y_j} - \frac{1 - \delta_{(j)}}{1 - y_j} \right\} \leq 0. \tag{1.13}$$

Thus $\tilde{y}$ maximizes (1.7).

Conversely, suppose that $\tilde{y}$ maximizes (1.7) under the constraint (1.8). If $0 < \epsilon < 1 - y_n$, the vector $\tilde{y} + \epsilon \tilde{1}_i$ satisfies (1.8), for each i, $1 \leq i \leq n$ (here we use $y_n < 1$). Since $\tilde{y}$ maximizes (1.7) we must have:

$$\lim_{\epsilon \downarrow 0} \frac{\phi(\tilde{y} + \epsilon \tilde{1}_i) - \phi(\tilde{y})}{\epsilon} = \sum_{j \geq i} \left\{ \frac{\delta_{(j)}}{y_j} - \frac{1 - \delta_{(j)}}{1 - y_j} \right\} \leq 0, \ 1 \leq i \leq n.$$

This yields (1.9).

Again using $y_1 > 0$ and $y_n < 1$, we obtain (1.10) by observing that

$$\lim_{h \to 0} \frac{\phi(\tilde{y} + h\tilde{y}) - \phi(\tilde{y})}{h} = 0,$$

since ϕ is differentiable at each $\tilde{y}$ such that $y_1 > 0$ and $y_n < 1$, and since the derivative of $h \mapsto \phi(\tilde{y} + h\tilde{y})$ is zero at $h = 0$, if ϕ attains a maximum at $\tilde{y}$.

Since ϕ is concave and, moreover, the (diagonal) matrix of second derivatives is non-singular at $\tilde{y}$, it follows that the vector $\tilde{y}$, satisfying (1.9) and (1.10), is the *unique* maximizing vector.   □

Remark 1.2. Proposition 1.1 is actually a special case of Fenchel's duality theorem, see e.g., Rockafellar (1970), Theorem 31.4. Since things simplify somewhat in the present setting, we chose to write down a proof of this special case.

Example 1.1. Let $n = 5$, $\delta_{(1)} = \delta_{(3)} = \delta_{(4)} = 1$, and $\delta_{(2)} = \delta_{(5)} = 0$. Then the vector $\tilde{y}$, maximizing (1.7) is given by

$$y_1 = y_2 = \tfrac{1}{2}, \ y_3 = y_4 = y_5 = \tfrac{2}{3}.$$

Note that y_m, $1 \leq m \leq 5$, is given by

$$y_m = \max_{i \leq m} \min_{k \geq m} \frac{\sum_{i \leq j \leq k} \delta_{(j)}}{k - i + 1}.$$

This is the so-called "max-min formula" for the solution of the maximization problem.

The solution can be found graphically by plotting the points $\left(i, \sum_{j \leq i} \delta_{(j)} \right)$ in the plane, and drawing the *(greatest) convex minorant* of these points on the interval $[0, 5]$. The convex minorant is defined as the function $H^* : [0, 5] \to \mathbb{R}$ such that

$$H^*(t) = \sup\{ \quad H(t) : H(i) \le \sum_{j \le i} \delta_{(j)}, \text{for each } i, 0 \le i \le 5,$$
$$H(0) = 0, \text{ and } H \text{ is convex}\},$$

for $t \in [0, 5]$, where $\sum_{j \le i} \delta_{(j)} \overset{\text{def}}{=} 0$, if $i = 0$. Then y_i is the *left derivative* of H^* at i, for $i = 1, \ldots, 5$.

The next proposition shows that the representation of $\tilde{y}$, given in Example 1.1, holds generally.

Proposition 1.2. Let H^* be the convex minorant of the points $\left(i, \sum_{j \le i} \delta_{(j)}\right)$ on $[0, n]$, i.e.,

$$H^*(t) = \sup\{ \quad H(t) : H(i) \le \sum_{j \le i} \delta_{(j)}, \text{for each } i, 0 \le i \le n,$$
$$H(0) = 0, \text{ and } H \text{ is convex}\},$$

for $t \in [0, n]$. Moreover, let y_i be the left derivative of H^* at i. Then $\tilde{y} = (y_1, \ldots, y_n)$ is the unique vector maximizing (1.7) under the constraint (1.8).

Remark 1.3. Note that in Proposition 1.2 (in contrast to Proposition 1.1) no restriction is made on $\delta_{(1)}$ and $\delta_{(n)}$.

Proof of Proposition 1.2. We start by first looking at some special (trivial) cases. If $\delta(i) = 0$, for each i, then $H^*(t) \equiv 0$, and hence $y_i = 0$, for each i. If $\delta(i) = 1$, for each i, then $H^*(t) = t$, $t \in [0, n]$, and hence $y_i = 1$, for each i. Finally, if $\delta_{(i)} = 0, 1 \le i \le k$ and $\delta_{(i)} = 1, k < i \le n$, for some $k < n$, then $y_i = 0, 1 \le i \le k$ and $y_i = 1, k < i \le n$. It is clear that in all these cases $\tilde{y}$ maximizes (1.7).

So we suppose that there is at least one $\delta_{(i)} = 1$, followed by a $\delta_{(j)} = 0$, for some $j > i$. Let k_0 be the smallest index i such that $\delta_{(i)} = 1$, and let m_0 be the largest index i such that $\delta_{(i)} = 0$. Then $y_i = 0$, if $i < k_0$, $y_i \in (0, 1)$, if $k_0 \le i \le m_0$, and $y_i = 1$, if $i > m_0$. Let k_1 be the largest index $k > k_0$ such that

$$\frac{\sum_{k_0 \le j < k} \delta_{(j)}}{k - k_0} = \min_{m > k_0} \frac{\sum_{k_0 \le j < m} \delta_{(j)}}{m - k_0}.$$

Then

$$y_i = y_{k_0} = \frac{\sum_{k_0 \le j < k_1} \delta_{(j)}}{k_1 - k_0}, \quad k_0 \le i < k_1,$$

and

$$\frac{\sum_{k_0 \le j < k} \delta_{(j)}}{k - k_0} \ge y_{k_0}, \quad k > k_0.$$

Hence we get, for each k, $k_0 < k \le k_1$:

$$\frac{\sum_{k_0 \le j < k}(1 - \delta_{(j)})}{\sum_{k_0 \le j < k} \delta_{(j)}} = \frac{k - k_0}{\sum_{k_0 \le j < k} \delta_{(j)}} - 1 \le \frac{1}{y_{k_0}} - 1 = \frac{1 - y_{k_0}}{y_{k_0}},$$

implying

$$\sum_{k_0 \leq j < k} \frac{\delta_{(j)}}{y_{k_0}} \Big/ \sum_{k_0 \leq j < k} \frac{1 - \delta_{(j)}}{1 - y_{k_0}} = \frac{1 - y_{k_0}}{y_{k_0}} \Big/ \frac{\sum_{k_0 \leq j < k}(1 - \delta_{(j)})}{\sum_{k_0 \leq j < k} \delta_{(j)}} \geq 1. \quad (1.14)$$

Thus:

$$\sum_{k_0 \leq j < k} \left\{ \frac{\delta_{(j)}}{y_{k_0}} - \frac{1 - \delta_{(j)}}{1 - y_{k_0}} \right\} \geq 0, \ k_0 < k \leq k_1. \quad (1.15)$$

If $k = k_1$, we obtain equality in (1.14) (and hence in (1.15)), since $\sum_{k_0 \leq j < k_1} \delta_{(j)} \Big/ (k_1 - k_0)$ is equal to y_{k_0}.

We can now repeat this argument for indices $i > k_1$: by definition

$$y_{k_1} = \max_{i \leq k_1} \min_{k \geq k_1} \frac{\sum_{i \leq j \leq k} \delta_{(j)}}{k - i + 1}.$$

Suppose k_2 is the largest index k such that

$$\frac{\sum_{k_1 \leq j < k} \delta_{(j)}}{k - k_1} = \max_{i \leq k_1} \min_{m \geq k_1} \frac{\sum_{i \leq j \leq m} \delta_{(j)}}{m - i + 1}.$$

Then we have, for $k_1 < k \leq k_2$:

$$\frac{\sum_{k_1 \leq j < k}(1 - \delta_{(j)})}{\sum_{k_1 \leq j < k} \delta_{(j)}} = \frac{k - k_1}{\sum_{k_1 \leq j < k} \delta_{(j)}} - 1 \leq \frac{1}{y_{k_1}} - 1 = \frac{1 - y_{k_1}}{y_{k_1}},$$

implying

$$\sum_{k_1 \leq j < k} \left\{ \frac{\delta_{(j)}}{y_{k_1}} - \frac{1 - \delta_{(j)}}{1 - y_{k_1}} \right\} \geq 0, k_1 < k \leq k_2, \quad (1.16)$$

(compare with (1.15)). If $k = k_2$, we obtain equality in (1.16).

Generally we get for a "block" of consecutive indices $k_m, \ldots, k_{m+1} - 1$ on which the function $i \mapsto y_i$ is constant:

$$\sum_{k_m \leq j < k} \left\{ \frac{\delta_{(j)}}{y_j} - \frac{1 - \delta_{(j)}}{1 - y_j} \right\} \geq 0, \ k_m < k \leq k_{m+1}, \quad (1.17)$$

with equality if $k = k_{m+1}$. Summing these (in)equalities over the successive blocks yields

$$\sum_{k_0 \leq j < m_0 + 1} \left\{ \frac{\delta_{(j)}}{y_j} - \frac{1 - \delta_{(j)}}{1 - y_j} \right\} = 0,$$

and, using (1.17):

$$\sum_{i \leq j \leq m_0} \left\{ \frac{\delta_{(j)}}{y_j} - \frac{1 - \delta_{(j)}}{1 - y_j} \right\} \leq 0, \ i = k_0, \ldots, m_0. \quad (1.18)$$

Moreover, since $i \mapsto y_i$ is constant on blocks of consecutive indices $k_m, \ldots, k_{m+1}-1$ and

$$\sum_{k_m \leq j < k_{m+1}} \left\{ \frac{\delta_{(j)}}{y_j} - \frac{1 - \delta_{(j)}}{1 - y_j} \right\} = 0, \tag{1.19}$$

on each block, we obtain

$$\sum_{i=k_0}^{m_0} \left\{ \frac{\delta_{(i)}}{y_i} - \frac{1 - \delta_{(i)}}{1 - y_i} \right\} y_i = 0. \tag{1.20}$$

But (1.18) and (1.20) correspond to (1.9) and (1.10), with the index 1 replaced by k_0 and the index n replaced by m_0.

The uniqueness of $\tilde{y}$ now easily follows from Proposition 1.1.   □

Proposition 1.2 shows that $i \mapsto y_i$ is the *isotonic regression* of the function $i \mapsto \delta_{(i)}$ in the class of all isotonic functions $i \mapsto x_i$, with respect to the simple ordering $x_1 \leq \ldots \leq x_n$. This means that the function $i \mapsto y_i$ minimizes

$$\sum_{i=1}^{n} \left\{ \delta_{(i)} - x_i \right\}^2,$$

in the class of such isotonic functions $i \mapsto x_i$, see, e.g., Theorem 1.2.1 on p. 7 of Robertson et al. (1988), where the connection between the derivative of the convex minorant and the solution of the isotonic regression problem is given. So we get that the NPMLE of F_0, maximizing (1.1), is given by

$$\hat{F}_n\big(T_{(i)}\big) = y_i,$$

where $\tilde{y}$ is the isotonic regression of the function $i \mapsto \delta_{(i)}$. Another (easier) route to this result is given in Exercise 5 at the end of this chapter. The method of proof, based on Proposition 1.1, has the advantage of being more generally applicable, as will be shown below.

We now turn to interval censoring, Case 2. Again we have two approaches, either based on the self-consistency equations or on the theory of isotonic regression. We start by deriving the self-consistency equations. The argument is similar to the argument, used to derive (1.5). The log likelihood, divided by n, can now be written in the following way:

$$\psi(F) \stackrel{\text{def}}{=} \int_{\mathbb{R}^3} \phi_F(x, t, u) \, dP_n(x, t, u), \tag{1.21}$$

where

$$\phi_F(x, t, u) \stackrel{\text{def}}{=} 1_{\{x \leq t\}} \log F(t) + 1_{\{t < x \leq u\}} \log\{F(u) - F(t)\} + 1_{\{x > u\}} \log\{1 - F(u)\},$$

and P_n is the empirical probability measure of the triples (X_i, T_i, U_i), $1 \leq i \leq n$.

Defining, for $t_0 \in \mathbb{R}$, the function $A_{t_0} : \mathbb{R} \to \mathbb{R}$ by

$$A_{t_0}(t) = \begin{cases} 0, & \text{if } F(t_0) = 0, \, t \in \mathbb{R}, \\ F(t \wedge t_0)/F(t_0) - F(t), & \text{if } F(t_0) > 0, \, t \in \mathbb{R}, \end{cases}$$

we get:

$$\lim_{h \downarrow 0} \frac{\psi(F + hA_{t_0}) - \psi(F)}{h}$$

$$= \int \Bigg\{ 1_{\{x \leq t\}} \frac{F(t \wedge t_0)/F(t_0) - F(t)}{F(t)}$$

$$+ 1_{\{t < x \leq u\}} \frac{\big(F(u \wedge t_0) - F(t \wedge t_0)\big)/F(t_0) - \big(F(u) - F(t)\big)}{F(u) - F(t)}$$

$$+ 1_{\{x > u\}} \frac{\big(1 - F(u \wedge t_0)/F(t_0)\big) - \big(1 - F(u)\big)}{1 - F(u)} \Bigg\} \, dP_n(x, t, u).$$

If F is the NPMLE, then this limit should be zero, for each t_0. This leads to the following equation for the NPMLE:

$$\begin{aligned} &\hat{F}_n(t_0) \\ &= \int \Bigg\{ 1_{\{x \leq t\}} \frac{\hat{F}_n(t \wedge t_0)}{\hat{F}_n(t)} + 1_{\{t < x \leq u\}} \frac{\hat{F}_n(u \wedge t_0) - \hat{F}_n(t \wedge t_0)}{\hat{F}_n(u) - \hat{F}_n(t)} \\ &\qquad\qquad\qquad + 1_{\{x > u\}} \frac{\hat{F}_n(t_0) - \hat{F}_n(u \wedge t_0)}{1 - \hat{F}_n(u)} \Bigg\} \, dP_n(x, t, u), \end{aligned} \tag{1.22}$$

which can also be written

$$\hat{F}_n(t) = E_{\hat{F}_n} \big\{ \, F_n(t) \, \big| \, T_1, \ldots, T_n, U_1, \ldots, U_n, \gamma_1, \ldots, \gamma_n, \delta_1, \ldots, \delta_n \, \big\}, \tag{1.23}$$

where F_n is the empirical distribution function of the random variables $X_1, \ldots, X_n$ (compare with (1.6)). So, in analogy with (1.6), we get that $\hat{F}_n(t_0)$ is the conditional expectation of the empirical distribution function F_n at t_0, given the available information $T_1, \ldots, T_n, U_1, \ldots, U_n$ and $\gamma_1, \ldots, \gamma_n, \delta_1, \ldots, \delta_n$ under the (self-induced) probability measure $P_{\hat{F}_n}$. Equation (1.23) yields the iteration steps of the EM algorithm, as will be shown in Chapter 3. Again the "self-consistency equation" (1.23) does not uniquely determine the NPMLE $\hat{F}_n$, even under the conventions of Remark 1.1.

We now turn to the isotonic regression approach. Let $\mathcal{F}$ be the class of distribution functions F satisfying

$$\begin{cases} F(T_i) > 0 & , \text{ if } X_i \leq T_i, \\ F(U_i) - F(T_i) > 0 & , \text{ if } T_i < X_i \leq U_i, \\ 1 - F(U_i) > 0 & , \text{ if } X_i > U_i, \end{cases} \tag{1.24}$$

and having mass concentrated on the set of observation points augmented with an extra point bigger than all observation points (see Remark 1.1). Note that if F maximizes the log likelihood (1.21), then F has to satisfy (1.24), since otherwise

$$\psi(F) = -\infty.$$

For distribution functions $F \in \mathcal{F}$, we define the process $t \mapsto W_F(t)$ by

$$
\begin{aligned}
W_F(t) = & \int_{t' \in [0,t],\, x \leq t'} F(t')^{-1} dP_n(x, t', u) \\
& - \int_{t' \in [0,t],\, t' < x \leq u} \{F(u) - F(t')\}^{-1} dP_n(x, t', u) \\
& + \int_{u \in [0,t],\, t' < x \leq u} \{F(u) - F(t')\}^{-1} dP_n(x, t', u) \\
& - \int_{u \in [0,t],\, x > u} \{1 - F(u)\}^{-1} dP_n(x, t', u), \\
& \hspace{6cm} \text{for } t \geq 0,
\end{aligned}
\tag{1.25}
$$

where P_n is the empirical probability measure of the points (X_i, T_i, U_i), $i = 1, \ldots, n$. The process W_F can only have a jump at an observation point t which is either a T_i or a U_i, and is such that the corresponding X_i either belongs to the interval to the right or to the interval to the left of this observation point. If, for example, $X_i \leq T_i$, then we will meet no terms of the form

$$
\frac{1}{n(F(U_i) - F(T_i))} \quad \text{or} \quad \frac{1}{n(1 - F(U_i))}
$$

in the process (1.25). This corresponds to the fact that there will be no terms of the form

$$
\log\{F(U_i) - F(T_i)\} \quad \text{or} \quad \log\{1 - F(U_i)\}
$$

in the log likelihood function. Likewise, if $\{\ X_i > U_i\ \}$, we will meet no terms of the form

$$
\frac{1}{nF(T_i)} \quad \text{or} \quad \frac{1}{n(F(U_i) - F(T_i))}
$$

in the process (1.25).

So we can "throw away" these irrelevant observation points. For later convenience, we will denote this "thinned" set of observation points by J_n, as is expressed by the following definition.

Definition 1.1. Let $J_n^{(1)}$ be the set of observation times T_i such that X_i either belongs to $[0, T_i]$ or to $(T_i, U_i]$, and let $J_n^{(2)}$ be the set of observation times U_i such that X_i either belongs to $(T_i, U_i]$ or to (U_i, ∞). Furthermore, let $J_n = J_n^{(1)} \cup J_n^{(2)}$, and let $T_{(j)}$ be the jth order statistic of the set J_n.

In the maximization problem we may assume, without loss of generality, that $T_{(1)}$ corresponds to an observation point T_i such that $\{X_i \leq T_i\} = 1$, and, similarly, we may assume that the largest order statistic in J_n, say $T_{(m)}$, corresponds to an observation point U_i such that $\{X_i > U_i\} = 1$. The reasons for this are similar to those discussed before Proposition 1.1. If, for example, $T_{(1)}$ would correspond to an observation point T_i such that $\{T_i < X_i \leq U_i\} = 1$, the distribution function F, maximizing (1.21), should satisfy $F(T_{(1)}) = 0$, since this makes the term $n^{-1} \log\{F(U_i) - F(T_i)\}$ as big as possible, without putting additional constraints on F. Similarly, if, for example, the largest order statistic $T_{(m)} \in J_n$ would correspond to an observation time U_i such that $\{T_i < X_i \leq U_i\} = 1$, then the maximizing F should satisfy $F(U_i) = 1$. In this case we can redefine the observation time T_i to be a right endpoint U_i' of an interval such that $\{X_i > U_i'\} = 1$. Since the left endpoint of such an interval would not belong to the set J_n, we get an equivalent maximization problem, with the left endpoint T_i replaced by a right endpoint U_i'. Finally, if the largest order statistic $T_{(m)} \in J_n$ would correspond to an observation time T_i such that $\{X_i \leq T_i\} = 1$, then the maximizing F satisfies $F(T_{(m)}) = 1$, and we can just remove this observation point from the set J_n, without altering the maximization problem.

We now get the following proposition, analogous to Proposition 1.1.

Proposition 1.3. Let $T_{(1)}$ correspond to an observation point T_i such that $\{X_i \leq T_i\} = 1$, and let the largest order statistic $T_{(m)} \in J_n$ correspond to an observation point U_i such that $\{X_i > U_i\} = 1$. Then $\hat{F}_n$ maximizes (1.21) over all $F \in \mathcal{F}$ if and only if

$$\int_{[t,\infty)} dW_{\hat{F}_n}(t') \leq 0, \quad \forall t \geq 0, \tag{1.26}$$

and

$$\int \hat{F}_n(t) \, dW_{\hat{F}_n}(t) = 0, \tag{1.27}$$

where W_F is defined by (1.24). Moreover, $\hat{F}_n$ is uniquely determined by (1.26) and (1.27).

Proof. Suppose $\hat{F}_n$ satisfies (1.26) and (1.27). Then, for all $F \in \mathcal{F}$,

$$\psi(F) - \psi(\hat{F}_n) \leq \int \left(F(t) - \hat{F}_n(t) \right) dW_{\hat{F}_n}(t).$$

This is shown in a similar way as (1.11) in the proof of Proposition 1.1. In fact, defining the function ϕ on the set

$$S = \left\{ \tilde{x} \in (0,1)^m : \tilde{x} = \left(F(T_{(1)}), \ldots, F(T_{(m)}) \right), \text{for some } F \in \mathcal{F} \right\}$$

by

$$\phi(\tilde{x}) = \psi(F), \text{if } \tilde{x} = \left(F(T_{(1)}), \ldots, F(T_{(m)}) \right),$$

it is seen that the maximization problem boils down to the problem of maximizing $\phi(\tilde{x})$ over the set S. Moreover, if $\tilde{x} = \left(F(T_{(1)}), \ldots, F(T_{(m)}) \right)$, we have

$$\frac{\partial}{\partial x_i} \phi(\tilde{x}) = W_F(T_{(i)}) - W_F(T_{(i-1)}), \, i = 1, \ldots, m,$$

where $T_{(0)} \stackrel{\text{def}}{=} 0$. Hence, if $\tilde{y} = \left(\hat{F}_n(T_{(1)}), \ldots, \hat{F}_n(T_{(m)}) \right)$, we get, since ϕ is concave

$$
\begin{aligned}
\psi(F) - \psi(\hat{F}_n) &= \phi(\tilde{x}) - \phi(\tilde{y}) \le \langle \nabla \phi(\tilde{y}), \tilde{x} - \tilde{y} \rangle \\
&= \int \left(F(t) - \hat{F}_n(t) \right) dW_{\hat{F}_n}(t) = \int F(t) \, dW_{\hat{F}_n}(t),
\end{aligned}
$$

using (1.27) in the last equality. Since $\tilde{x}$ can be represented as

$$
\tilde{x} = \sum_{i=1}^{m} \alpha_i \tilde{1}_i,
$$

where α_i and $\tilde{1}_i$ are defined as in (1.12), we get

$$
\int F(t) \, dW_{\hat{F}_n}(t) = \sum_{i=1}^{m} \alpha_i \langle \nabla \phi(\tilde{y}), \tilde{1}_i \rangle = \sum_{i=1}^{m} \alpha_i \int_{[T_{(i)}, \infty)} dW_{\hat{F}_n}(t) \le 0.
$$

Conversely, suppose that $\tilde{y} = \left(\hat{F}_n(T_{(1)}), \ldots, \hat{F}_n(T_{(m)}) \right)$ and that $\hat{F}_n$ maximizes (1.21) over $\mathcal{F}$. Then

$$
\lim_{\epsilon \downarrow 0} \frac{\phi(\tilde{y} + \epsilon \tilde{1}_i) - \phi(\tilde{y})}{\epsilon} = \int_{[T_{(m-i+1)}, \infty)} dW_{\hat{F}_n}(t) \le 0 \quad, 1 \le i \le m.
$$

and

$$
\lim_{h \to 0} \frac{\phi(\tilde{y} + h\tilde{y}) - \phi(\tilde{y})}{h} = \int \hat{F}_n(t) \, dW_{\hat{F}_n}(t) = 0.
$$

Finally, we get by a Taylor expansion with a Lagrangian remainder term

$$
\begin{aligned}
&\phi(\tilde{x}) - \phi(\tilde{y}) - \langle \nabla \phi(\tilde{y}), \tilde{x} - \tilde{y} \rangle \\
&= -\frac{1}{2n} \sum_{i=1}^{n} p_i \bigg\{ \left(F(T_i) - \hat{F}_n(T_i) \right)^2 \{ X_i \le T_i \} \\
&\qquad + \left\{ F(U_i) - F(T_i) - \left(\hat{F}_n(U_i) - \hat{F}_n(T_i) \right) \right\}^2 \{ T_i < X_i \le U_i \} \\
&\qquad + \left(F(U_i) - \hat{F}_n(U_i) \right)^2 \{ X_i > U_i \} \bigg\},
\end{aligned} \tag{1.28}
$$

where $p_i > 0$, $1 \le i \le n$. If, for example $\{ X_i \le T_i \} = 1$, then $p_i = 1/z_i^2$, where z_i is a point in the open interval with endpoints x_i and y_i, if $x_i \ne y_i$, and where $z_i = y_i$, if $x_i = y_i$. The values $\hat{F}_n(T_{(i)})$ are "linked" to each other in terms of the form $n^{-1} \log \{ \hat{F}_n(U_i) - \hat{F}_n(T_i) \}$ and $n^{-1} \log \{ \hat{F}_n(U_j) - \hat{F}_n(T_j) \}$, with $U_i < T_j$ and $\hat{F}_n(U_i) = \hat{F}_n(T_j)$ (see Exercise 7 and Examples 1.2 and 1.3 below). Moreover, by (1.28) and the assumptions $\{ X_i \le T_{(1)} \} = 1$ and $\{ X_j > T_{(m)} \} = 1$ for some indices i and j, we have $F(T_{(1)}) = \hat{F}_n(T_{(1)})$ and $F(T_{(m)}) = \hat{F}_n(T_{(m)})$. From this it is easily checked that the right-hand side of (1.28) can only be zero if $F(T_{(i)}) = \hat{F}_n(T_{(i)})$, for $i = 1, \ldots, m$. This proves the uniqueness. $\square$

Example 1.2. Consider the following maximization problem. Maximize

$$\phi(x_1, \ldots, x_5) = \log x_1 + \log(x_4 - x_2) + \log(1 - x_3) + \log(1 - x_5),$$

under the restriction

$$0 \le x_1 \le \ldots \le x_5 \le 1.$$

This corresponds to an interval censoring problem, case 2, where $n = 4$ and

$$X_1 \le T_1, \; T_2 < X_2 \le U_2, \; X_3 > U_3, \; X_4 > U_4,$$
$$T_{(1)} = T_1, \; T_{(2)} = T_2, \; T_{(3)} = U_3, \; T_{(4)} = U_2, \text{ and } T_{(5)} = U_4.$$

Let

$$\tilde{y} = (y_1, \ldots, y_5) = \left(\hat{F}_n(T_{(1)}), \ldots, \hat{F}_n(T_{(5)}) \right).$$

Then (1.27) yields:

$$2 - \frac{y_3}{1 - y_3} - \frac{y_5}{1 - y_5} = 0,$$

and (1.26) yields:

$$-\frac{1}{1 - y_5} \le 0, \quad \frac{1}{y_4 - y_2} - \frac{1}{1 - y_5} \le 0, \quad -\frac{1}{1 - y_3} + \frac{1}{y_4 - y_2} - \frac{1}{1 - y_5} \le 0,$$
$$-\frac{1}{1 - y_3} - \frac{1}{1 - y_5} \le 0, \quad \frac{1}{y_1} - \frac{1}{1 - y_3} - \frac{1}{1 - y_5} \le 0.$$

These equations and inequalities are satisfied for

$$y_1 = y_2 = y_3 = \tfrac{1}{4}, \; y_4 = y_5 = \tfrac{5}{8}.$$

The following example shows that the solution vector may have irrational components.

Example 1.3. Consider maximizing

$$\begin{aligned}
\phi(x_1, \ldots, x_{12}) \;=\; & \log x_1 + \log(x_4 - x_2) + \log(1 - x_3) + \log x_5 \\
& + \log(1 - x_6) + \log x_8 + \log(1 - x_9) + \log x_{10} \\
& + \log(x_{11} - x_7) + \log(1 - x_{12}),
\end{aligned}$$

under the restriction

$$0 \le x_1 \le \ldots \le x_{12} \le 1.$$

This would correspond to an interval censoring problem with $n = 10$ and

$$\delta_i = 1, \text{if } i = 1, 4, 7 \text{ and } 9,$$
$$\gamma_i = 1, \text{if } i = 2 \text{ and } 6,$$
$$\gamma_i = \delta_i = 0, \text{ otherwise.}$$

The set of (relevant) observation times is ordered as follows.

$$T_1 < T_2 < U_3 < U_2 < T_4 < U_5 < T_6 < T_7 < U_8 < T_9 < U_6 < U_{10}.$$

In this case the NPMLE $\hat{F}_n$ (with $n = 10$) and the maximizing vector $\tilde{y}$ are given by

$$\hat{F}_n(T_{(i)}) = y_i = \begin{cases} \frac{1}{2} - \frac{1}{6}\sqrt{3}, & 1 \leq i \leq 3, \\ \frac{1}{2}, & 4 \leq i \leq 9, \\ \frac{1}{2} + \frac{1}{6}\sqrt{3}, & 10 \leq i \leq 12. \end{cases}$$

We will now give a characterization of the NPMLE as the left derivative of the convex minorant of a cumulative sum diagram with "self-induced weights". We start by introducing the processes on which this cumulative sum diagram is based.

Let the processes G_F and V_F be defined by

$$\begin{aligned} G_F(t) = \ & \int_{t' \in [0,t],\, x \leq t'} F(t')^{-2} dP_n(x, t', u) \\ & + \int_{t' \in [0,t],\, t' < x \leq u} \{F(u) - F(t')\}^{-2} dP_n(x, t', u) \\ & + \int_{u \in [0,t],\, t' < x \leq u} \{F(u) - F(t')\}^{-2} dP_n(x, t', u), \\ & + \int_{u \in [0,t],\, x > u} \{1 - F(u)\}^{-2} dP_n(x, t', u), \\ & \hspace{6cm} \text{for } t \geq 0, \end{aligned} \tag{1.29}$$

and

$$V_F(t) = W_F(t) + \int_{[0,t]} F(t') \, dG_F(t'), \, t \geq 0. \tag{1.30}$$

The following proposition characterizes the NPMLE $\hat{F}_n$ as the slope of the convex minorant of a self-induced cumulative sum diagram.

Proposition 1.4. Let $T_{(1)}$ correspond to an observation point T_i such that $\{X_i \leq T_i\} = 1$, and let the largest order statistic $T_{(m)} \in J_n$ correspond to an observation point U_i such that $\{X_i > U_i\} = 1$. Then $\hat{F}_n$ is the NPMLE of F_0 if and only if $\hat{F}_n$ is the left derivative of the convex minorant of the "cumulative sum diagram", consisting of the points

$$P_j = \left(G_{\hat{F}_n}(T_{(j)}), V_{\hat{F}_n}(T_{(j)}) \right),$$

where $P_0 = (0,0)$ and $T_{(j)} \in J_n$, $j = 1, 2, \ldots$

Proof. For simplicity of notation, we will write V_n, W_n and G_n instead of $V_{\hat{F}_n}$, $W_{\hat{F}_n}$ and $G_{\hat{F}_n}$, respectively.

By definition, the left derivative h_n of the convex minorant of the cumulative sum diagram is given by

$$h_n(\tau_i) = \frac{V_n(\tau_i) - V_n(\tau_{i-1})}{G_n(\tau_i) - G_n(\tau_{i-1})}$$

at the successive locations τ_i of the vertices of the convex minorant of the cumulative sum diagram. Moreover, defining $T_{(0)} = 0$ and

$$\Delta V_{n,i} = V_n(T_{(i)}) - V_n(T_{(i-1)}), \text{ and } \Delta G_{n,i} = G_n(T_{(i)}) - G_n(T_{(i-1)}),$$

we have that h_n minimizes

$$\sum_{T_{(i)} \in J_n} \left\{ h(T_{(i)}) - \frac{\Delta V_{n,i}}{\Delta G_{n,i}} \right\}^2 \Delta G_{n,i}$$

over all nondecreasing functions h, such that $h(0) = 0$. This means by Theorem 1.5, p. 28 in Barlow et al. (1972) or Theorem 1.3.2 in Robertson et al. (1988) that

$$\sum_i \left\{ \frac{\Delta V_{n,i}}{\Delta G_{n,i}} - h_n(T_{(i)}) \right\} h(T_{(i)}) \Delta G_{n,i} \leq 0, \tag{1.31}$$

and

$$\sum_i \left\{ \frac{\Delta V_{n,i}}{\Delta G_{n,i}} - h_n(T_{(i)}) \right\} h_n(T_{(i)}) \Delta G_{n,i} = 0, \tag{1.32}$$

for all nondecreasing h, such that $h(0) = 0$. But it is easily verified that (1.31) implies (1.26) and that (1.32) implies (1.27), with $\hat{F}_n$ replaced by h_n. Proposition 1.3 now implies that h_n is the NPMLE.

Conversely, if $\hat{F}_n$ satisfies (1.26) and (1.27), then $\hat{F}_n$ also satisfies (1.31) and (1.32), with h_n replaced by $\hat{F}_n$. This in turn implies that $\hat{F}_n$ is the left derivative of the convex minorant of the cumulative sum diagram.   □

We will see in Chapter 3 that Proposition 1.4 actually leads to an iterative convex minorant algorithm for computing the NPMLE, which seems to converge much faster than the EM algorithm.

Remark 1.4. We note that the "weight function" G_F could also be chosen in another way, still yielding the same characterization of the NPMLE as the left derivative of the convex minorant of a self-induced cumulative sum diagram. The particular choice of G_F, made here, is based on a second order expansion of the log likelihood function. We will return to this point in Chapter 3.

1.2 Exercises

1. Show that the value of the left derivative of the function H^* in Proposition 1.2 is given by

$$h^*(m) = \max_{i \leq m} \min_{k \geq m} \frac{\sum_{i \leq j \leq k} \delta_{(j)}}{k - i + 1}.$$

2. Show that the following definitions of h^* are equivalent to the definition, given in Exercise 1:

$$h^*(m) = \min_{k \geq m} \max_{i \leq m} \frac{\sum_{i \leq j \leq k} \delta_{(j)}}{k - i + 1}$$

$$h^*(m) = \max_{i \leq m} \min_{k \geq i} \frac{\sum_{i \leq j \leq k} \delta_{(j)}}{k - i + 1}$$

$$h^*(m) = \min_{k \geq m} \max_{i \leq k} \frac{\sum_{i \leq j \leq k} \delta_{(j)}}{k - i + 1}.$$

3. Show that the function ϕ, defined in Example 1.3, is maximized by taking

$$x_i = \begin{cases} \frac{1}{2} - \frac{1}{6}\sqrt{3}, & 1 \leq i \leq 3, \\ \frac{1}{2}, & 4 \leq i \leq 9, \\ \frac{1}{2} + \frac{1}{6}\sqrt{3}, & 10 \leq i \leq 12. \end{cases}$$

4. Can condition (1.26) in Proposition 1.3 be replaced by the following condition?

$$\int_{[t,\infty)} \hat{F}_n(u)\, dW_{\hat{F}_n}(u) \leq 0, \quad \forall t \geq 0. \tag{1.26*}$$

5. Can condition (1.27) in Proposition 1.3 be replaced by the following condition?

$$\int dW_{\hat{F}_n}(t) = 0. \tag{1.27*}$$

(The following exercise gives an alternative proof of Proposition 1.2.)

6. Let the function $H : [0, n] \rightarrow \mathbb{R}$ be defined by

$$H(i) = \sum_{j=1}^{i} \delta_{(j)}, \ 1 \leq i \leq n, \ H(0) = 0, \tag{E.1}$$

where $\delta_{(j)}$ is defined as in Proposition 1.2, and where H is defined by linear interpolation at values $x \in [0, n] \backslash \{0, \ldots, n\}$. We want to maximize the expression

$$\int_{[0,n]} \log y(x)\, dH(x) + \int_{[0,n]} \log\{1 - y(x)\}\, d(x - H(x)),$$

over all nondecreasing functions $y : [0, n] \rightarrow [0, 1]$, such that

$$y(x) = y(i), \ x \in (i - 1, i], \ 1 \leq i \leq n, \ y(0) = 0. \tag{E.2}$$

Let H^* be the convex minorant of H on $[0, n]$, i.e.,

$$H^*(x) = \max\{f(x) : f : [0, n] \rightarrow \mathbb{R} \text{ is convex and } f \leq H\}, \ x \in [0, n].$$

(a) Show that

$$\int_{[0,n]} \log y(x)\, dH(x) \leq \int_{[0,n]} \log y(x)\, dH^*(x), \qquad (E.3)$$

and

$$\int_{[0,n]} \log\{1 - y(x)\}\, d(x - H(x)) \leq \int_{[0,n]} \log\{1 - y(x)\}\, d(x - H^*(x)), \qquad (E.4)$$

for all nondecreasing functions $y : [0, n] \to [0, 1]$, satisfying (E.2) (defining $0 \cdot (-\infty) = 0$, $a \cdot (-\infty) = -\infty$, if $a > 0$).

(b) Let h^* be the left derivative of H^*. Show that

$$\int_{[0,n]} \log h^*(x)\, dH(x) + \int_{[0,n]} \log\{1 - h^*(x)\}\, d(x - H(x))$$
$$= \int_{[0,n]} \log h^*(x)\, dH^*(x) + \int_{[0,n]} \log\{1 - h^*(x)\}\, d(x - H^*(x)).$$

(c) Show that

$$\int_{[0,n]} \log\{h(x)/h^*(x)\}\, dH^*(x) + \int_{[0,n]} \log\{(1-h(x))/(1-h^*(x))\}\, d(x - H^*(x)) \leq 0,$$

for each non-decreasing function $h : [0, n] \to [0, 1]$.

(d) Deduce Proposition 1.2 from (a), (b) and (c).

7. Show that the function ϕ, defined by

$$\phi(\tilde{x}) = \psi(F), \text{if } \tilde{x} = \left(F(T_{(1)}), \ldots, F(T_{(m)})\right),$$

in the proof of Proposition 1.3, is generally not *strictly* concave. Show that one can always make a preliminary reduction, "linking" components of the vector $\tilde{x}$, such that we get a strictly concave function on a (convex subset of a) space of lower dimension. For example, in Example 1.2, we can immediately make the following reduction:

$$x_1 = x_2 = x_3; \quad x_4 = x_5.$$

So in this case the function ϕ reduces to a function $\bar{\phi}$ of two variables:

$$\bar{\phi}(x_1, x_2) = \log x_1 + \log(x_2 - x_1) + \log(1 - x_1) + \log(1 - x_2).$$

The function $\bar{\phi}$ is strictly concave on the set $\{ (x_1, x_2) : 0 < x_1 < x_2 < 1 \}$.

2 The Deconvolution Problem

2.1 Decreasing densities and non-negative random variables

We first consider the deconvolution problem in a model with non-negative random variables and disturbances with a decreasing density. Formally, let $Z_1, \ldots, Z_n$ be a sample from a distribution function H with density

$$h(z) = \int g(z - x) \, dF_0(x), \ z \in \mathbb{R}, \tag{2.1}$$

where g is a decreasing density on $[0, \infty)$, and F_0 an unknown distribution function, concentrated on $[0, \infty)$. For example, g could be the exponential density

$$g(x) = e^{-x} 1_{[0,\infty)}(x), \ x \in \mathbb{R},$$

or the Uniform (0,1) density

$$g(x) = 1_{[0,1]}(x), \ x \in \mathbb{R}.$$

An NPMLE of F_0 is a distribution function, maximizing

$$\psi(F) = \int \log \left\{ \int g(z - x) \, dF(x) \right\} dH_n(z), \tag{2.2}$$

as a function of F, where H_n is the empirical distribution function of the sample $Z_1, \ldots, Z_n$.

In order to define the self-consistency equations, we proceed as before. Defining the direction A_t as in (1.4), we find

$$\lim_{h \downarrow 0} \frac{\psi(F + hA_t) - \psi(F)}{h} = \int \frac{\int g(z - x) \, dA_t(x)}{\int g(z - x) \, dF(x)} \, dH_n(z). \tag{2.3}$$

Putting (2.3) equal to zero yields the following equation for the NPMLE $\hat{F}_n$:

$$\begin{aligned} \hat{F}_n(t) &= \int \frac{\int_{(-\infty, t]} g(z - x) \, d\hat{F}_n(x)}{\int g(z - x) \, d\hat{F}_n(x)} \, dH_n(z) \\ &= n^{-1} \sum_{i=1}^{n} P_{\hat{F}_n} \left\{ X_i \leq t \mid Z_1, \ldots, Z_n \right\}. \end{aligned} \tag{2.4}$$

Letting $Z_i = X_i + Y_i$, where $X_1, \ldots, X_n$ and $Y_1, \ldots, Y_n$ are independent samples from distributions with distribution function F_0 and density g, respectively, we can write (2.4) as

$$\hat{F}_n(t) = E_{\hat{F}_n} \left\{ F_n(t) \mid Z_1, \ldots, Z_n \right\}, \tag{2.5}$$

where F_n is the empirical distribution function of $X_1, \ldots, X_n$. So we again get a self-consistency equation, similar to (1.6) (again not uniquely determining $\hat{F}_n$).

Turning to the isotonic regression approach, we first note that we may assume that a distribution function F, maximizing (2.2), is a discrete distribution function, with mass concentrated at the points $Z_1, \ldots, Z_n$. Let $Z_{(1)}, \ldots, Z_{(n)}$ be the order statistics of the sample $Z_1, \ldots, Z_n$. Then we may also assume that the maximizing F satisfies $F(Z_{(1)}) > 0$ and $F(Z_{(n)}) = 1$ (see Exercise 1 of this chapter). In analogy with (1.25), we introduce a process $t \mapsto W_F(t)$, defined by

$$W_F(t) = \sum_{Z_{(i)} \leq t} \Delta W_F(Z_{(i)}), \tag{2.6}$$

where

$$\begin{aligned}
\Delta W_F(Z_{(i)}) &= \int \frac{g(z - Z_{(i)}) - g(z - Z_{(i+1)})}{\int g(z - x)\, dF(x)}\, dH_n(z), \ 1 \leq i < n, \\
\Delta W_F(Z_{(n)}) &= \int \frac{g(z - Z_{(n)})}{\int g(z - x)\, dF(x)}\, dH_n(z).
\end{aligned} \tag{2.7}$$

Let $\mathcal{F}$ be the class of discrete distribution functions, with masses concentrated at the points $Z_1, \ldots, Z_n$, and satisfying $F(Z_{(1)}) > 0$. We get the following proposition, analogous to Proposition 1.3.

Proposition 2.1. The distribution function $\hat{F}_n \in \mathcal{F}$ maximizes (2.2) over $\mathcal{F}$ if and only if

$$\int_{[0,t]} dW_{\hat{F}_n}(t') \geq 0, \quad \forall t \geq 0, \tag{2.8}$$

and

$$\int \{\hat{F}_n(t) - 1\}\, dW_{\hat{F}_n}(t) = 0, \tag{2.9}$$

where W_F is defined by (2.6). Moreover, $\hat{F}_n \in \mathcal{F}$ is uniquely determined by (2.8) and (2.9).

Proof. We proceed in a similar way as in the proof of Proposition 1.3. Define the function ϕ on the set

$$S = \left\{ \tilde{x} \in [0,1)^n : \tilde{x} = \left(1 - F(Z_{(n)}), \ldots, 1 - F(Z_{(1)})\right), \text{ for some } F \in \mathcal{F} \right\}$$

by

$$\phi(\tilde{x}) = \psi(F), \text{ if } \tilde{x}_i = 1 - F(Z_{(n-i+1)}), \ 1 \leq i \leq n.$$

Then

$$\frac{\partial}{\partial x_i} \phi(\tilde{x}) = -\Delta W_F(Z_{(n-i+1)}), \ i = 1, \ldots, n,$$

where $\Delta W_F(Z_{(i)})$ is defined by (2.7). Note that ϕ is well defined, since, by definition, $F(Z_{(1)}) > 0$, if $F \in \mathcal{F}$.

Hence, if $\tilde{y} = \left(1 - \hat{F}_n(Z_{(n)}), \ldots, 1 - \hat{F}_n(Z_{(1)})\right)$, we get, since ϕ is concave

$$\psi(F) - \psi(\hat{F}_n) = \phi(\tilde{x}) - \phi(\tilde{y}) \leq \langle \nabla \phi(\tilde{y}), \tilde{x} - \tilde{y} \rangle$$
$$= \int \left\{ (1 - \hat{F}_n(t)) - (1 - F(t)) \right\} dW_{\hat{F}_n}(t) = -\int \left(1 - F(t)\right) dW_{\hat{F}_n}(t) \leq 0,$$

using (2.9) in the last equality, and (2.8) in the last inequality. The remaining part of the proof is analogous to the proof of Proposition 1.3, except for the proof of the uniqueness.

For the proof of the uniqueness, let A be the matrix of second derivatives of the function ϕ at $\tilde{x}$, where $\tilde{x} = \left(1 - F(Z_{(n)}), \ldots, 1 - F(Z_{(1)})\right)$, for some $F \in \mathcal{F}$.

Then we get, if $\tilde{u} \in I\!\!R^n$:

$$\tilde{u}'A\tilde{u} = -n^{-1} \sum_{j=1}^{n} \beta_j^{-2} \left\{ \sum_{i \leq j} \alpha_{ij} u_{n-i+1} \right\}^2, \tag{2.10}$$

where

$$\beta_j = \int g(Z_{(j)} - x)\, dF(x),$$
$$\alpha_{ji} = \alpha_{ij} = g(Z_{(j)} - Z_{(i)}) - g(Z_{(j)} - Z_{(i+1)}),\ i < j \leq n,$$

and

$$\alpha_{ii} = g(0),\ 1 \leq i \leq n.$$

It is seen from (2.10) that $\tilde{u}'A\tilde{u}$ can only be zero if $\tilde{u} = 0$. Hence A is non-singular, implying that ϕ is strictly concave. $\square$

Remark 2.1. The reason for considering the vector $\tilde{x} = \left(1 - F(Z_{(n)}), \ldots, 1 - F(Z_{(1)})\right)$ in the proof of Proposition 2.1, instead of the vector $\left(F(Z_{(1)}), \ldots, F(Z_{(n)})\right)$, is the fact that we then only have to consider a maximization problem over the whole *cone* $\{\tilde{x} : 0 \leq x_1 \leq \ldots \leq x_n\}$ instead of a maximization problem over the more complicated bounded convex region $\{\tilde{x} : 0 < x_1 \leq \ldots \leq x_n \leq 1\}$ (defining the function to be $-\infty$ if $x_{(n)} \geq 1$).

We also note that the NPMLE $\hat{F}_n$ always satisfies $\hat{F}_n(Z_{(n)}) = 1$, but that we do not have to build this into the conditions of Proposition 2.1.

Although in some cases the NPMLE can be found by a 1-step procedure, just by computing the slope of the convex minorant of a certain function (see Exercises 2 and 3 for deconvolution with, respectively, the uniform and the exponential densities as convolution kernels), this does not seem to be possible in general. However, we can (as in the case of interval censoring) give a characterization of the NPMLE as the left derivative of the convex minorant of a cumulative sum diagram with self-induced weights. This leads to an iterative convex minorant algorithm, which (in our experience) converges rather fast and will be discussed in Chapter 3.

Let the process G_F be defined by

$$G_F(t) = \sum_{Z_{(i)} \leq t} \Delta G_F(Z_{(i)}), \tag{2.11}$$

where

$$
\begin{aligned}
\Delta G_F(Z_{(i)}) &= \int \frac{\left\{g(z - Z_{(i)}) - g(z - Z_{(i+1)})\right\}^2}{\left\{\int g(z - x)\, dF(x)\right\}^2}\, dH_n(z),\ 1 \leq i < n, \\
\Delta G_F(Z_{(n)}) &= \int \frac{g(z - Z_{(n)})^2}{\left\{\int g(z - x)\, dF(x)\right\}^2}\, dH_n(z),
\end{aligned} \tag{2.12}
$$

and let the process V_F be defined by

$$V_F(t) = W_F(t) + \int_{[0,t]} F(t')\, dG_F(t'),\ t \geq 0. \tag{2.13}$$

The following proposition, analogous to Proposition 1.4, characterizes the NPMLE $\hat{F}_n$ as the slope of the convex minorant of a "self-induced" cumulative sum diagram.

Proposition 2.2. The distribution function $\hat{F}_n \in \mathcal{F}$ maximizes (2.2) over $\mathcal{F}$ if and only if $\hat{F}_n$ satisfies

$$\hat{F}_n(T_{(i)}) = c_n(T_{(i)}) \wedge 1,\ 1 \leq i \leq n,$$

where c_n is the left derivative of the convex minorant of the "cumulative sum diagram", consisting of the points

$$P_j = \left(G_{\hat{F}_n}(Z_{(j)}), V_{\hat{F}_n}(Z_{(j)})\right),\ 0 \leq j \leq n,$$

and where $P_0 = (0,0)$.

The proof is similar to the proof of Proposition 1.4 and is therefore omitted. The process G_F contains "second derivatives on the diagonal" of the log likelihood function, but has again a certain arbitrariness in the sense that other weight functions would give the same type of characterization in Proposition 2.2. However, computer experiments show a superiority of this weight function with respect to certain other possibilities.

Remark 2.2. The results of this section can be generalized to the situation where g is non-decreasing on an interval which may be different from $[0, \infty)$ and where F_0 need not be zero on $(-\infty, 0)$, but we will not go into this here.

2.2 Convolution with symmetric densities

The deconvolution problem with symmetric kernels offers some new features, which even make the computation of the NPMLE more difficult. The most important difference is that, unlike in the situation discussed in section 2.1, we generally cannot assume that the NPMLE corresponds to a probability distribution, with mass concentrated on the observation points.

We will assume that the convolution kernel g satisfies the following conditions:

$$g \text{ is symmetric about the origin, i.e., } g(x) = g(-x), \ x \in \mathbb{R} \qquad (2.14)$$

and

$$g \text{ is continuous and decreases on } \mathbb{R}_+. \qquad (2.15)$$

Examples of such densities are: the standard normal, the Cauchy, the Laplace (or double exponential) and the triangular densities. Again we assume that the sample of "observables" is $Z_1, \ldots, Z_n$ where Z_i has density

$$h(z) = \int g(z - x) \, dF_0(x), \ z \in \mathbb{R}, \qquad (2.16)$$

There is a voluminous literature on this problem and on methods for estimating the density f_0, corresponding to the distribution function F_0, see, e.g., Carroll and Hall (1988), Fan (1988), Stefanski and Carroll (1987) and Zhang (1990). In these papers the estimation method is invariably based on Fourier inversion. Almost nothing seems to be known about the behavior of the NPMLE (we will, however, establish consistency of the NPMLE in Chapter 4).

The self-consistency equation (2.5) applies without change, and can actually be used in the EM algorithm. However, the convergence of the EM algorithm is in this case so painfully slow that it almost seems useless for practical purposes. Below some information about the general structure of the NPMLE is listed. The line of argument is similar to arguments used by Jewell (1982) in characterizing the NPMLE of the mixing distribution in scale mixtures of exponential distributions.

It will first be shown that an NPMLE (i.e., a distribution function $\hat{F}_n$, maximizing $\psi(F)$, defined by (2.2), with g satisfying (2.14) and (2.15)) always exists.

Lemma 2.1. There always exists an NPMLE. Moreover, the vector

$$\left(\int g(Z_{(1)} - x) \, d\hat{F}_n(x), \ldots, \ \int g(Z_{(n)} - x) \, d\hat{F}_n(x) \right)$$

has the same value for each NPMLE $\hat{F}_n$.

Proof. Let the function $\tilde{x} \mapsto k(\tilde{x})$, $\tilde{x} \in [0, 1]^n$ be defined by

$$k(\tilde{x}) = \begin{cases} n^{-1} \sum_{i=1}^{n} \log x_i & , \ x_i \in (0, 1], \ 1 \le i \le n, \\ -\infty & , \ \text{if } x_i = 0, \text{ for some } i. \end{cases}$$

Moreover, let $\mathcal{F}$ be the set of subdistribution functions on $\mathbb{R}$, and let the function $\chi : \mathcal{F} \to [0, 1]^n$ be defined by

$$\chi(F) = \left(\int g(Z_{(1)} - x) \, dF(x), \ldots, \ \int g(Z_{(n)} - x) \, dF(x) \right), \ F \in \mathcal{F}.$$

The set $M = \{\chi(F) : F \in \mathcal{F}\}$ is convex and compact (in the vague topology) and hence there exists a subdistribution function $\bar{F} \in \mathcal{F}$ such that

$$k\big(\chi(\bar{F})\big) = \sup_{\tilde{x} \in M} k(\tilde{x}).$$

Since the set $\{\chi(F) : F \in \mathcal{F},\ \chi(F) \in (0,1]^n\}$ is nonempty, we may assume $\chi(\bar{F}) \in (0,1]^n$, since otherwise $k\big(\chi(\bar{F})\big) = -\infty$. Moreover, since k is strictly concave on $(0,1]^n$, there is a *unique* $\tilde{y} \in (0,1]^n$ such that

$$k(\tilde{y}) = \sup_{F \in \mathcal{F}} k\big(\chi(F)\big) = \sup_{\tilde{x} \in M} k(\tilde{x}).$$

Finally, if $\bar{F}$ maximizes $k\big(\chi(F)\big)$, then $\bar{F}$ has to be a distribution function, since otherwise there would exist an $\epsilon > 0$ such that

$$k\Big(\chi\big(\bar{F} + \epsilon 1_{[0,\infty)}\big)\Big) > k\big(\chi(\bar{F})\big).$$

□

The following proposition yields some information about the set on which the NPMLE is concentrated.

Proposition 2.3. Let $P_{\hat{F}_n}$ the probability measure on $\mathbb{R}$, corresponding to an NPMLE $\hat{F}_n$, and let the set M be defined by

$$M = \left\{ y \in \mathbb{R} : \int \left\{ g(z-y) \Big/ \int g(z-x)\, d\hat{F}_n(x) \right\} dH_n(z) = 1 \right\}.$$

Then $P_{\hat{F}_n}(M) = 1$. Moreover,

$$\int \frac{g(z-y)}{\int g(z-x)\, d\hat{F}_n(x)}\, dH_n(z) < 1, \text{if } y \notin M.$$

Proof. Fix $x \in \mathbb{R}$. Then

$$\lim_{h \downarrow 0} h^{-1} \left\{ \psi\big(\hat{F}_n + h(1_{[x,\infty)} - \hat{F}_n)\big) - \psi(\hat{F}_n) \right\}$$

$$= \int \frac{g(z-x) - \int g(z-y)\, d\hat{F}_n(y)}{\int g(z-y)\, d\hat{F}_n(y)}\, dH_n(z) \qquad (2.17)$$

$$= \int \frac{g(z-x)}{\int g(z-y)\, d\hat{F}_n(y)}\, dH_n(z) - 1 \le 0,$$

since $\hat{F}_n$ is an NPMLE. By integrating with respect to the measure $P_{\hat{F}_n}$ (and by Fubini's theorem) we get

$$\int_{x \in \mathbb{R}} \left\{ \int_{z \in \mathbb{R}} \frac{g(z-x)}{\int g(z-y)\, d\hat{F}_n(y)}\, dH_n(z) \right\} d\hat{F}_n(x) = 1. \qquad (2.18)$$

So we must have $P_{\hat{F}_n}(M) = 1$, since otherwise, by (2.17), the left side of (2.18) would be strictly smaller than 1. □

Proposition 2.3 shows that in many cases the support of $P_{\hat{F}_n}$ is a finite set. For example, if g is a normal density we get the following result.

Corollary 2.1 Let g be a normal density, symmetric about zero. Then $P_{\hat{F}_n}$ is concentrated on a finite set of at most n points.

Proof. Without loss of generality we may assume that g is the standard normal density. Proposition 2.3 implies

$$\int \left\{ \exp\{-\tfrac{1}{2}(z-y)^2\} \Big/ \int \exp\{-\tfrac{1}{2}(z-x)^2\} \, d\hat{F}_n(x) \right\} dH_n(z) = 1,$$

if $y \in M$. This means that y satisfies an equation of the form

$$\sum_{i=1}^{n} \alpha_i \exp\{-\tfrac{1}{2}(Z_{(i)} - y)^2\} - 1 = 0, \tag{2.19}$$

with $\alpha_i > 0$, $1 \le i \le n$. By Karlin and Studden (1966), pp. 9-11, Examples 1 and 5, this equation can have at most n roots. □

Remark 2.3. It is clear that Corollary 2.1 also holds for normal densities which are not symmetric about zero.

As a consequence of the following corollary, we get a stronger result in the case of the double exponential density.

Corollary 2.2 Let g be strictly convex on $[0, \infty)$. Then $P_{\hat{F}_n}$ is concentrated on the set of observation points.

Proof. Let $\alpha_i = 1 \big/ \int g(Z_{(i)} - x) \, d\hat{F}_n(x)$, $1 \le i \le n$. Then the function

$$x \mapsto n^{-1} \sum_{i=1}^{n} \alpha_i g(x - Z_{(i)}), \; x \in \mathbb{R},$$

is strictly convex on each interval $(Z_{(i-1)}, Z_{(i)})$, $1 \le i \le n+1$, where $Z_{(0)} = -\infty$ and $Z_{(n+1)} = \infty$. Thus, if the set M is defined as in Proposition 2.3, we get:

$$x \in M \implies x = Z_i, \text{ for some } i, \; 1 \le i \le n,$$

since

$$x \notin \{Z_1, \ldots, Z_n\} \implies n^{-1} \sum_{i=1}^{n} \alpha_i g(x - Z_{(i)}) < 1.$$

□

The following Corollary of Proposition 2.3 can also be obtained by direct methods.

Corollary 2.3 Let g satisfy

$$g(x) < g(0), \text{ for all } x \neq 0.$$

Then $P_{\hat{F}_n}\left[Z_{(1)}, Z_{(n)}\right] = 1$, i.e., $P_{\hat{F}_n}$ is concentrated on the range of the data points.

Example 2.1. (The following curious facts have been communicated to me by Rudolf Grübel.) Let g be the standard normal density. We want to find the support of $\hat{F}_n$ if the sample size is $n = 2$. After a suitable shift we may assume $Z_{(1)} = -a$, $Z_{(2)} = a$. We have:

$$\psi(F) = \tfrac{1}{2}\left\{\log \int \phi(a - x)\, dF(x) + \log \int \phi(a + x)\, dF(x)\right\}.$$

If $\tilde{F}$ is defined by $\tilde{F}(x) = 1 - F(-x-)$, where

$$F(-x-) = \lim_{y\uparrow -x,\, y<-x} F(y),$$

we get

$$\psi(\tilde{F}) = \psi(F).$$

Moreover, since ψ is concave, we have

$$\psi\left(\tfrac{1}{2}(F + \tilde{F})\right) \geq \tfrac{1}{2}\{\psi(F) + \psi(\tilde{F})\}.$$

Hence, if $\psi(F)$ is maximal, then $\psi(F^s)$ is also maximal, where F^s is the symmetrized distribution function $F^s = \tfrac{1}{2}\{F + \tilde{F}\}$.

So, in order to find an NPMLE, we may restrict ourselves to symmetric distributions F. For these we have

$$\int_{(-\infty,0)} \phi(-a - x)\, dF(x) = \int_{(0,\infty)} \phi(-a + x)\, dF(x),$$

and likewise

$$\int_{(-\infty,0)} \phi(a - x)\, dF(x) = \int_{(0,\infty)} \phi(a + x)\, dF(x),$$

implying (by the symmetry of ϕ):

$$\psi(F) = \log\left\{\phi(a)P_F(\{0\}) + \int_{(0,\infty)} \{\phi(a - x) + \phi(a + x)\}\}\, dF(x)\right\}.$$

Since, by Proposition 2.3, the support of a maximum likelihood estimator consists of at most 2 points, we either get the degenerate distribution at zero, or a (symmetric) distribution concentrated on the two points y and $-y$, where $y > 0$ satisfies

$$\phi(a - y) + \phi(a + y) = \sup_{x>0}\{\phi(a - x) + \phi(a + x)\}.$$

We have:

$$\frac{d}{dy}\{\phi(a-y)+\phi(a+y)\}=0 \iff \frac{a-y}{a+y}=\frac{\phi(a+x)}{\phi(a-x)}.$$

For $u = y/a$ this yields

$$\frac{1-u}{1+u}=\exp\{-2a^2 u\}. \tag{2.20}$$

As a function of u, the left hand side of (2.20) is strictly decreasing to -1, as $u \to \infty$, and the right hand side of (2.20) tends to 0 in this case. Both are differentiable with strictly increasing derivatives, and take the value 1 at $u = 0$. Hence another point of intersection exists if and only if the derivative of the left hand side is strictly smaller than the corresponding quantity for the right hand side, i.e., if and only if $a > 1$.

Putting all this together, we get that generally the NPMLE will be the degenerate distribution at the mean $\bar{z} = \frac{1}{2}\{z_1 + z_2\}$ of the two observations if $a = \frac{1}{2}\{z_{(2)} - z_{(1)}\} \leq 1$ and will "split" into two masses of size $1/2$ each, located at the points $\bar{z} \pm ac(a)$, where $c(a) = u$ is the (unique) positive solution of (2.20). Some values of $c(a)$ are:

$$c(1.01) = 0.24121, \ c(1.1) = 0.66993, \ c(2) = 0.99933.$$

A similar situation arises in the Cauchy and analogous cases. Generally, the "bifurcation distance" of the two data points will be $2u$, where u separates the convex and concave part of the symmetric density g.

2.3 Exercises

1. Show that, under the conditions of section 2.1, there exists an NPMLE $\hat{F}_n$ with mass concentrated on the set of observation points, and satisfying $\hat{F}_n(Z_{(1)}) > 0$ and $\hat{F}_n(Z_{(n)}) = 1$. Consider in particular the case $g = 1_{[0,1]}$ (i.e., g is the Uniform $(0,1)$ density). Is the NPMLE always uniquely determined, if we drop the condition that its mass is concentrated on the set of observation points?

2. Let g be the Uniform $(0,1)$ density, and suppose that we know that the unknown distribution function F_0 is also concentrated on $[0,1]$. Defining $\delta_i = 1_{\{Z_i \leq 1\}}$, the function ψ, defined by (2.2), can be written

$$\psi(F) = n^{-1}\sum_{i=1}^{n}\Big\{\delta_i \log F(Z_i) + (1-\delta_i)\log\big(1-F(Z_i - 1)\big)\Big\}.$$

(a) Let $Y_i, 1 \leq i \leq n$, be defined by

$$Y_i = \begin{cases} Z_i & , \text{ if } \delta_i = 1, \\ Z_i - 1 & , \text{ if } \delta_i = 0. \end{cases}$$

Show that $Y_1, \ldots, Y_n$ is distributed as a sample from a Uniform $(0,1)$ distribution.

(b) Let $Y_{(1)}, \ldots, Y_{(n)}$ be the set of order statistics of the set $Y_1, \ldots, Y_n$, and define

$$\delta_{(i)} = \begin{cases} 1 & , \text{ if the } X_k, \text{ corresponding to } Y_{(i)}, \text{ is } \leq 1, \\ 0 & , \text{ otherwise.} \end{cases}$$

Show that the value at $Y_{(i)}$ of the NPMLE $\hat{F}_n$, maximizing $\psi(F)$ over $\mathcal{F}$, is given by the left continuous derivative at the point i of the convex minorant of the function $K_n : [0, n] \to \mathbb{R}$, defined by

$$K_n(i) = \sum_{j \leq i} \delta_{(j)}, \ K_n(0) = 0,$$

at the points i, and by linear interpolation at other points of $[0, n]$.

3. (This example is due to R. Grübel. For a different approach, see Vardi (1989).) Let g be the exponential density

$$g(x) = e^{-x} 1_{[0,\infty)}(x), \ x \in \mathbb{R},$$

and let the function ψ be defined by (2.2). Show that the value at $Z_{(i)}$ of the NPMLE $\hat{F}_n$, maximizing $\psi(F)$ over $\mathcal{F}$ can be found as follows.

Let the points P_i, $0 \leq i \leq n$, be defined by

$$P_i = (x_i, y_i) = \begin{cases} (0,0) & , \text{ if } i = 0, \\ \left(e^{-Z_{(1)}} - e^{-Z_{(i+1)}}, i/n\right) & , \text{ if } 1 \leq i < n, \\ \left(e^{-Z_{(1)}}, 1\right), & , \text{ if } i = n. \end{cases}$$

Let f_n be the left derivative on $(0, x_n]$ of the convex minorant of the cumulative sum diagram, consisting of the points P_i, and let $f_n(0) = 0$. Then the value at $Z_{(i)}$ of the NPMLE $\hat{F}_n$ is given by

$$\hat{F}_n(Z_{(i)}) = \sum_{j \leq i} e^{-Z_{(j)}} \left\{ f_n(x_j) - f_n(x_{j-1}) \right\}.$$

Hints: Since f_n is the left derivative of the convex minorant of the cumulative sum diagram, consisting of the points P_i, the function $i \mapsto f_n(x_i)$ is the isotonic regression of $u_i \stackrel{\text{def}}{=} 1/\{n(x_i - x_{i-1})\}$, $1 \leq i \leq n$, with weights $\Delta x_i = x_i - x_{i-1}$, i.e., this function minimizes

$$\sum_{i=1}^{n} \{f(i) - u_i\}^2 \Delta x_i,$$

over all functions $i \mapsto f(i)$, which are nondecreasing in i. This implies, by e.g., Robertson et al. (1988), Theorem 1.3.2:

$$\sum_{i=1}^{n} f_n(x_i)\{u_i - f_n(x_i)\}\Delta x_i = 0,$$

and

$$\sum_{j \geq i} \{u_j - f_n(x_j)\} \Delta x_j \leq 0, \ 1 \leq i \leq n.$$

Show that the conditions of Proposition 2.3 follow from these relations.

4. Deduce from Karlin and Studden (1966), pp. 9-11, Examples 1 and 5, that equation (2.19) can have at most n roots.

5. Let g be the triangular density

$$g(x) = (1 - |x|) 1_{[-1,1]}(x), \ x \in \mathbb{R}.$$

Can we assume (as in the double exponential case) that the mass of the NPMLE is concentrated on the set of data points?

3 Algorithms

3.1 The EM algorithm

We illustrate the properties of the EM algorithm by studying its behavior in the interval censoring problem, Case 1. As argued in Remark 1.1 of Chapter 1 of part II, we may, in our search for the NPMLE, restrict attention to a class $\mathcal{F}$ of purely discrete distribution functions F with mass concentrated on the set of points $T_{(1)}, \ldots, T_{(n)}, T_{(n+1)}$, where $T_{(n+1)}$ is an arbitrary point $t > T_{(n)}$ (we need a point to put the remaining mass on). Let

$$p_i = P_F\{X = T_{(i)}\}, \ 1 \leq i \leq n+1.$$

The EM algorithm runs as follows. We take a starting distribution $P_{F^{(0)}}$, with positive masses at all points $T_{(i)}, 1 \leq i \leq n+1$. For example, we could take the discrete uniform distribution

$$p_i^{(0)} = P_{F^{(0)}}\{X = T_{(i)}\} = 1/(n+1), \ 1 \leq i \leq n+1. \tag{3.1}$$

We then do the "E"-step ("E" for "Expectation"), i.e., we compute the conditional expectation of the log likelihood

$$E^{(0)}\left\{\sum_{i=1}^{n} \log f(X_i) \mid \delta_1, \ldots, \delta_n, T_1, \ldots, T_n\right\}, \tag{3.2}$$

where $f(x) = P_F\{X = x\}$ and $E^{(0)}$ is the expectation under the probability measure $P_{F^{(0)}}$. Next we maximize (3.2) over all discrete distribution with probability density f with respect to counting measure on the set $\{T_{(1)}, \ldots, T_{(n+1)}\}$. This yields new probability masses $p_i^{(1)}, \ i = 1, \ldots, n+1$.

Now we repeat the E and M step, starting with the probability distribution with masses $p_i^{(1)}$ instead of the $p_i^{(0)}$, defined by (3.1), etc.

In order to establish the connection with the self-consistency equations (1.6), we take a closer look at what happens in the E and M steps. We can rewrite (3.2) in the following way:

$$
\begin{aligned}
&E^{(0)}\left\{\sum_{i=1}^{n} \log f(X_i) \mid \delta_1, \ldots, \delta_n, T_1, \ldots, T_n\right\} \\
&= \sum_{i=1}^{n} E^{(0)}\left\{\log f(X_i) \mid \delta_1, \ldots, \delta_n, T_1, \ldots, T_n\right\} \\
&= \sum_{i=1}^{n}\left\{\sum_{k=1}^{n+1} (\log p_k) P^{(0)}\{X_i = T_{(k)} \mid \delta_i, T_i\}\right\} \\
&= \sum_{k=1}^{n+1} (\log p_k) \sum_{i=1}^{n} P^{(0)}\{X_i = T_{(k)} \mid \delta_i, T_i\},
\end{aligned}
\tag{3.3}
$$

where we write $P^{(0)}$ instead of $P_{F^{(0)}}$ (generally we will write $P^{(m)}$ instead of $P_{F^{(m)}}$). For the M-step, we have to maximize (3.3) over all discrete distributions with masses p_i at $T_{(i)}$. It is easily seen (for example by using Lagrangian multipliers) that (3.3) is maximized by taking

$$p_k = n^{-1} \sum_{i=1}^{n} P^{(0)}\big\{X_i = T_{(k)} \mid \delta_i, T_i\big\}, \ 1 \le k \le n+1.$$

Hence the combined E and M steps yield

$$p_k^{(1)} = n^{-1} \sum_{i=1}^{n} P^{(0)}\big\{X_i = T_{(k)} \mid \delta_i, T_i\big\}, \ 1 \le k \le n+1.$$

Generally, we get

$$p_k^{(m+1)} = n^{-1} \sum_{i=1}^{n} P^{(m)}\big\{X_i = T_{(k)} \mid \delta_i, T_i\big\}, \ 1 \le k \le n+1, \tag{3.4}$$

which leads to

$$F^{(m+1)}(t) = n^{-1} \sum_{i=1}^{n} P^{(m)}\big\{X_i \le t \mid \delta_i, T_i\big\} = E^{(m)}\big\{F_n(t) \mid \delta_1, \ldots, \delta_n, T_1, \ldots, T_n\big\},$$

where F_n is the empirical distribution function of the (unobservable) sample $X_1, \ldots, X_n$. This corresponds to the self-consistency equation (1.6).

One can write (3.4) in the following explicit form:

$$p_k^{(m+1)} \ = \ n^{-1} \sum_{i=1}^{n} \left\{ \frac{\delta_i}{F^{(m)}(T_i)} 1_{\{T_i \ge T_{(k)}\}} + \frac{1-\delta_i}{1 - F^{(m)}(T_i)} 1_{\{T_i < T_{(k)}\}} \right\} p_k^{(m)},$$
$$1 \le k \le n+1,$$

since

$$P^{(m)}\big\{X_i = T_{(k)} \mid \delta_i, T_i\big\}$$
$$= p_k^{(m)} \left\{ \frac{\delta_i}{F^{(m)}(T_i)} 1_{\{T_i \ge T_{(k)}\}} + \frac{1-\delta_i}{1 - F^{(m)}(T_i)} 1_{\{T_i < T_{(k)}\}} \right\}.$$

It follows that the probability masses $\hat{p}_k$, corresponding to the NPMLE, should satisfy

$$\hat{p}_k \ = \ n^{-1} \sum_{i=1}^{n} \left\{ \frac{\delta_i}{\hat{F}_n(T_i)} 1_{\{T_i \ge T_{(k)}\}} + \frac{1-\delta_i}{1 - \hat{F}_n(T_i)} 1_{\{T_i < T_{(k)}\}} \right\} \hat{p}_k,$$
$$1 \le k \le n+1.$$

Hence, if $\hat{p}_k > 0$, we get:

$$1 = n^{-1} \sum_{i=1}^{n} \left\{ \frac{\delta_i}{\hat{F}_n(T_i)} 1_{\{T_i \geq T_{(k)}\}} + \frac{1 - \delta_i}{1 - \hat{F}_n(T_i)} 1_{\{T_i < T_{(k)}\}} \right\}, \qquad (3.5)$$

and if $\hat{p}_k = 0$, we get

$$1 \geq n^{-1} \sum_{i=1}^{n} \left\{ \frac{\delta_i}{\hat{F}_n(T_i)} 1_{\{T_i \geq T_{(k)}\}} + \frac{1 - \delta_i}{1 - \hat{F}_n(T_i)} 1_{\{T_i < T_{(k)}\}} \right\}. \qquad (3.6)$$

Note that we do not necessarily get (3.5) *and* (3.6) *if we would not start with positive masses at all points* $T_{(k)}$! If the starting distribution puts zero masses at some points $T_{(k)}$, the EM algorithm would converge to a solution of the self-consistency equations, but this solution would not necessarily maximize the likelihood. This is caused by the fact that the EM algorithm will, at each iteration step, put mass zero at a point where the starting distribution puts mass zero. So, if the NPMLE puts positive mass at such a point, the EM algorithm will not converge to the NPMLE, if the starting distribution puts zero mass at that point.

There is the general empirical finding that the number of iteration steps, needed by the EM algorithm to get, say, the NPMLE in two accurate decimals, will increase with the sample size. Below we offer some speculations on why this will be true. In order to make things as simple as possible, we consider interval censoring, Case 1, and suppose that F_0 and G are both the uniform distribution function on $[0, 1]$.

The arguments in Wu (1983), which show that the EM algorithm (starting with positive weights at all points $T_{(1)}, \ldots, T_{(n+1)}$ will actually converge to the NPMLE, give no information about the speed of convergence and it is not clear that one has a contraction with a constant strictly smaller than 1. Moreover, it is not clear what norm should be used, if one wants to show that the iteration steps (3.4) really correspond to a contraction. We will now give some arguments which indicate that one indeed has a (local) contraction with respect to a certain L_2-norm (to be specified below), but with a constant which can be arbitrarily close to 1. By "local" we mean that the values of $F^{(m)}$ should not be too far from the values of the solution $\hat{F}_n$. We conjecture that one does not have a (local) contraction with respect to the supremum norm (on the space of distribution functions).

Let E be a linear space of bounded functions on $[0, 1]$, containing differences of (two) distribution functions, and let the mapping $\phi : E \to E$ be defined by

$$[\phi(h)](t) = h(t) - \int_{u \geq t} \frac{t}{u} h(u)\, du - \int_{u < t} \frac{1-t}{1-u} h(u)\, du \qquad (3.7)$$

The mapping ϕ can be considered as an approximate derivative of the mapping, defined by the iteration steps (3.4), since, for n *and* m large, we get

$$F^{(m+1)}(t) - F^{(m)}(t)$$
$$\approx \frac{1}{n} \sum_{T_i \geq t} \frac{\delta_i}{F^{(m)}(T_i)} \left(F^{(m)}(t) - F^{(m-1)}(t) \right)$$
$$+ \frac{1}{n} \sum_{T_i < t} \frac{1 - \delta_i}{1 - F^{(m)}(T_i)} \left(F^{(m)}(t) - F^{(m-1)}(t) \right)$$
$$- \frac{1}{n} \sum_{T_i \geq t} \frac{\delta_i}{F^{(m)}(T_i)^2} F^{(m)}(t) \left(F^{(m)}(T_i) - F^{(m-1)}(T_i) \right)$$
$$- \frac{1}{n} \sum_{T_i < t} \frac{1 - \delta_i}{\left(1 - F^{(m)}(T_i) \right)^2} \left(1 - F^{(m)}(t) \right) \left(F^{(m)}(T_i) - F^{(m-1)}(T_i) \right)$$

$$\approx F^{(m)}(t) - F^{(m-1)}(t) - \int_{u \geq t} \frac{t}{u} \left(F^{(m)}(u) - F^{(m-1)}(u) \right) du$$
$$- \int_{u < t} \frac{1-t}{1-u} \left(F^{(m)}(u) - F^{(m-1)}(u) \right) du,$$

where the approximations are (partly) justified by the (strong or weak) law of large numbers.

Continuing on this path (and leaving lots of problems aside), we obtain the following approximation to the squared L_2-distance between $F^{(m+1)}$ and $F^{(m)}$:

$$\| F^{(m+1)} - F^{(m)} \|^2$$
$$\approx \| F^{(m)} - F^{(m-1)} \|^2 - \frac{1}{2} \int \int_{[0,1]^2} \left\{ \frac{t \wedge u}{t \vee u} + \frac{1 - t \vee u}{1 - t \wedge u} \right\}$$
$$\cdot \left(F^{(m)}(t) - F^{(m-1)}(t) \right) \left(F^{(m)}(u) - F^{(m-1)}(u) \right) dt\, du,$$

where the L_2-distance is just the ordinary L_2-distance between (complex- or real-valued) functions on $[0,1]$ with respect to Lebesgue measure. So, in the limit (as the sample size tends to infinity), we have to deal with the mapping $f \mapsto Uf$, defined by

$$[Uf](t) = \frac{1}{2} \int_{[0,1]} \left\{ \frac{t \wedge u}{t \vee u} + \frac{1 - t \vee u}{1 - t \wedge u} \right\} f(u)\, du,$$

which is a compact self-adjoint operator on the pre-Hilbert space of continuous complex-valued functions f, defined on $[0,1]$. Note that the eigenvalues λ of U satisfy $0 \leq \lambda \leq 1$ (as is seen using the Cauchy-Schwarz inequality and the fact that U is self-adjoint). The operator U has eigenvalues arbitrarily close to zero.

On the other hand, for each n, the operator corresponds to a symmetric matrix with elements

$$\frac{1}{2n} \left\{ \frac{T_i \wedge T_j}{T_i \vee T_j} + \frac{1 - T_i \vee T_j}{1 - T_i \wedge T_j} \right\}.$$

For example, if $n = 2$ and $T_1 = 1/3$, $T_2 = 2/3$ we get the matrix

$$C_1 = \begin{pmatrix} \frac{1}{2} & \frac{1}{4} \\ \frac{1}{4} & \frac{1}{2} \end{pmatrix}$$

with eigenvalues $\frac{3}{4}$ and $\frac{1}{4}$, and if $n = 3$ and $T_i = i/4$ we get the matrix

$$
C_2 = \begin{pmatrix}
\frac{1}{3} & \frac{7}{36} & \frac{1}{9} \\[2mm]
\frac{7}{36} & \frac{1}{3} & \frac{7}{36} \\[2mm]
\frac{1}{9} & \frac{7}{36} & \frac{1}{3}
\end{pmatrix}
$$

The eigenvalues of the matrix $I - C_2$ are: 0.8916, 0.7778 and 0.3306.

This has a rather interesting general structure: the elements of the matrix C_i become smaller if one moves away from the diagonal, corresponding to a decreasing "coverage" of the points T_i and T_j by intervals $[0, T_k]$ and $(T_l, 1]$ in the estimation problem. One gets the impression that the biggest eigenvalue of the matrix $I - C_n$ will tend to 1, as $n \to \infty$, which would mean that the "contraction constant" also tends to 1, as $n \to \infty$. This would explain why the convergence of the EM algorithm becomes slower, if the sample size increases.

Remark 3.1. The following type of example suggests that we cannot expect to have a contraction with respect to the supremum norm. Let $f_\epsilon : [0, 1] \to \mathbb{R}$ be defined by

$$
f_\epsilon(x) = \begin{cases}
-1, & t \in \left[0, \frac{1}{2}\right), \\
1, & t \in \left[\frac{1}{2}, \frac{1}{2} + \epsilon\right), \\
0, & t \in \left[\frac{1}{2} + \epsilon, 1\right].
\end{cases}
$$

Then:

$$
\lim_{\epsilon \downarrow 0} [\phi f_\epsilon]\left(\tfrac{1}{2}\right) = 1 + \int_0^{\frac{1}{2}} \frac{1}{2(1 - u)} = 1 + \tfrac{1}{2} \log 2,
$$

where ϕ is defined by (3.7). So, for sufficiently small $\epsilon > 0$, we get $[\phi f_\epsilon](\frac{1}{2}) > 1$. This example can easily be changed into a similar example for matrices (showing that the infinite dimensionality of the function space is not essential here).

3.2 The iterative convex minorant algorithm

We discuss the properties of the iterative convex minorant algorithm for interval censoring, Case 2. The idea is to reduce the maximization problem to a series of weighted isotonic regression problems, where the weights are induced by the values of the solution at the preceding step.

Suppose $F^{(m)}$ is the estimator of F_0, found at the mth iteration step. Then $F^{(m+1)}$ is the distribution function, maximizing the expression

$$
\int \left(F(t) - F^{(m)}(t)\right) dW_{F^{(m)}}(t) - \tfrac{1}{2} \int \left(F(t) - F^{(m)}(t)\right)^2 dG_{F^{(m)}}(t), \qquad (3.8)
$$

as a function of $F \in \mathcal{F}$, where $W_{F^{(m)}}$ and $G_{F^{(m)}}$ are defined by (1.25) and (1.29), respectively, and $\mathcal{F}$ is the set of purely discrete distribution functions with mass concentrated on the set of points $T_{(i)}$ and possibly with mass at an extra point bigger than all observation points (see the beginning of the chapter and Remark 1.1 in Chapter 1). One can consider (3.8) as a second order approximation to $\psi(F) - \psi(F^{(m)})$, where the off-diagonal second order terms are omitted. It can be shown (Exercise 5) that the distribution function $F^{(m+1)}$, maximizing (3.8), is the left derivative of the convex minorant of the cumulative sum diagram, consisting of the points

$$P_j = \left(G_{F^{(m)}}(T_{(j)}), V_{F^{(m)}}(T_{(j)})\right),$$

where $P_0 = (0,0)$ and $T_{(j)} \in J_n$, $j = 1, 2, \ldots$, provided we make a preliminary reduction of the observation points $T_{(i)}$ such that the conditions of Proposition 1.4 are satisfied.

In the iteration steps of the algorithm we use "buffers" to prevent that $F^{(m)}$ would not be a distribution function or that the likelihood at $F^{(m)}$ would become zero. For example, one can always take a small positive number $c > 0$ such that $F^{(m)}(U_i) - F^{(m)}(T_i) \geq c$, for each pair (T_i, U_i) such that $\{T_i < X_i \leq U_i\} = 1$, since for $c > 0$ sufficiently small a term $\log\left(F^{(m)}(U_i) - F^{(m)}(T_i)\right) \leq \log c$ would make the log likelihood smaller than that of the discrete uniform distribution on the set of points (with the extra point bigger than all observation points). Likewise one can deal with values $F^{(m)}(T_i)$ or $1 - F^{(m)}(T_i)$ that would be too small.

In the full second order approximation to $\psi(F) - \psi(F^{(m)})$, we would have to add to (3.8) the term

$$\int_{t < x \leq u} \frac{\left(F(t) - F^{(m)}(t)\right)\left(F(u) - F^{(m)}(u)\right)}{\left(F^{(m)}(u) - F^{(m)}(t)\right)^2} \, dP_n(x, t, u). \tag{3.9}$$

By a simple change of the algorithm, one can actually maximize the full second order approximation to $\psi(F) - \psi(F^{(m)})$. In this modification, the "time scale" $G_{F^{(m)}}$ at step m is the same as in (1.29), but the process W_F is replaced by

$$\begin{aligned}
W_F^{(m)}(t) \\
= \int_{t' \in [0,t], \, x \leq t'} & \frac{2F^{(m)}(t') - F(t')}{F^{(m)}(t')^2} \, dP_n(x, t', u) \\
+ \int_{t' \in [0,t], \, t' < x \leq u} & \frac{F(u) - F(t') - 2\left\{F^{(m)}(u) - F^{(m)}(t')\right\}}{\left\{F^{(m)}(u) - F^{(m)}(t')\right\}^2} \, dP_n(x, t', u) \\
- \int_{u \in [0,t], \, t' < x \leq u} & \frac{F(u) - F(t') - 2\left\{F^{(m)}(u) - F^{(m)}(t')\right\}}{\left\{F^{(m)}(u) - F^{(m)}(t')\right\}^2} \, dP_n(x, t', u) \\
+ \int_{u \in [0,t], \, x > u} & \frac{1 - F(u) - 2\left\{1 - F^{(m)}(u)\right\}}{\left\{1 - F^{(m)}(u)\right\}^2} \, dP_n(x, t', u),
\end{aligned}$$

$$\text{for } t \geq 0,$$

and the process V_F is replaced by

$$V_F^{(m)}(t) = W_F^{(m)}(t) + \int_{[0,t]} F(t') \, dG_{F^{(m)}}(t'), \, t \geq 0.$$

So, within the mth iteration we have another sequence of iterations, where $F^{(m,i-1)}$ changes into $F^{(m,i)}$ at the ith step, and where $F^{(m,i)}$ is the left derivative of the convex minorant of the "cumulative sum diagram", consisting of the points

$$P_j = \left(G_{F^{(m)}}(T_{(j)}), V^{(m)}_{F^{(m,i-1)}}(T_{(j)}) \right),$$

with $P_0 = (0,0)$ and $T_{(j)} \in J_n$, $j = 1, 2, \ldots$ (see Definition 1.1 and Proposition 1.4).

By this modification, the "outer iterations" of the algorithm get Newton-type properties: it is more important to start close to the solution of the maximization problem, but once one is close, the convergence is very fast. However, the only similarity with a real Newton algorithm is that we maximize a second order approximation; *the crucial difference is that we hit the boundary of the parameter space at each step.* Only in the lower dimensional boundary set partial derivatives of the likelihood function become zero, but finding this lower dimensional boundary set is the main part of the problem!

If we define the L_2-distance $\| \cdot \|$ on $\mathcal{F}$ by

$$\|F_1 - F_2\|^2 = \int \left(F_1(t) - F_2(t) \right)^2 dG_{F^{(m)}}(t),$$

(suppressing the dependence on m in order to simplify notation), we have for the "inner" iteration steps of the modified algorithm:

$$\|F^{(m,i+1)} - F^{(m,i)}\| \leq \|F^{(m,i)} - F^{(m,i-1)}\|.$$

This can be shown as follows. Let the function

$$t \mapsto \frac{dV^{(m)}_F}{dG_{F^{(m)}}}(t)$$

be defined by

$$\frac{dV^{(m)}_F}{dG_{F^{(m)}}}(t) = \begin{cases} \dfrac{V^{(m)}_F(t) - V^{(m)}_F(t-)}{G_{F^{(m)}}(t) - G_{F^{(m)}}(t-)} & , \text{if } G_{F^{(m)}}(t) > G_{F^{(m)}}(t-), \\ 0 & , \text{otherwise.} \end{cases}$$

Then $F^{(m,i+1)}$ minimizes

$$\left\| F - \frac{dV^{(m)}_{F^{(m,i)}}}{dG_{F^{(m)}}} \right\|$$

as a function of F, where F is a purely discrete distribution function with support contained in the set of points $T_{(i)}$.

Let the mapping $T : \mathcal{F} \to \mathcal{F}$ be defined by

$$\left\| T(F) - \frac{dV^{(m)}_F}{dG_{F^{(m)}}} \right\| = \min_{H \in \mathcal{F}} \left\| H - \frac{dV^{(m)}_F}{dG_{F^{(m)}}} \right\|.$$

By Theorem 8.2.5 in Robertson et al (1988),

$$\left\| F^{(m,i+1)} - F^{(m,i)} \right\| = \left\| T\big(F^{(m,i)}\big) - T\big(F^{(m,i-1)}\big) \right\| \leq \left\| \frac{dV_{F^{(m,i)}}^{(m)}}{dG_{F^{(m)}}} - \frac{dV_{F^{(m,i-1)}}^{(m)}}{dG_{F^{(m)}}} \right\|.$$
(3.11)

But the square of the term at the right-hand side of (3.11) can be written

$$
\int_{t < x \leq u} \frac{\left\{ F^{(m,i)}(u) - F^{(m,i-1)}(u) \right\}^2}{\left\{ F^{(m)}(u) - F^{(m)}(t) \right\}^2} \, dP_n(x,t,u)
$$
$$
+ \int_{t < x \leq u} \frac{\left\{ F^{(m,i)}(t) - F^{(m,i-1)}(t) \right\}^2}{\left\{ F^{(m)}(u) - F^{(m)}(t) \right\}^2} \, dP_n(x,t,u)
$$
$$
\leq \left\| F^{(m,i)} - F^{(m,i-1)} \right\|^2
$$
(3.12)

Moreover, under the conditions of Proposition 1.4, there can only be equality in (3.12), if $F^{(m,i)} = F^{(m,i-1)}$. Hence by a subsequence argument and the fact that the mapping $T : \mathcal{F} \to \mathcal{F}$ has a unique fixed point $F^{(m,\infty)}$, $F^{(m,i)}$ converges to $F^{(m,\infty)}$ at each point $T_{(j)}$, as $i \to \infty$.

Our general experience with the two versions of the iterative convex minorant algorithm is that the simple version converges rather quickly from any starting point and that the modified version is considerable slower for large data set, mainly because the "inner iterations" converge so slowly at the beginning steps of the algorithm. So we have the peculiar situation that there are rather convincing arguments for the (at least local) convergence of the modified algorithm (the outer steps maximize the full second order approximation to the log likelihood and the inner steps can be shown to converge), but that the simple iterative convex minorant algorithm, maximizing only a diagonal approximation to the log likelihood at each step, shows much better convergence properties, although no convergence proof is available at the moment!

As indicated in Chapter 1, we can always make a preliminary reduction, linking values of the solution. A simple rule for doing this runs as follows. Consider the process V_F, defined by (1.30). Under the conditions of Proposition 1.3, the first jump ·of the process will be upward, and the last jump will be downward. We now form blocks of upward jumps followed by downward jumps. Specifically, if an upward jump is denoted by + and a downward jump by −, we consider consecutive blocks of the form

$$+ \ldots - \ldots,$$

where each block is made as big as possible. In Example 1.2 we would get the two consecutive blocks:

$$+ - -$$
$$+ -$$

and in Example 1.3 we would get the 4 consecutive blocks:

$$+ - -$$
$$+ + - -$$
$$+ -$$
$$+ - -$$

It is clear that the solution has to be constant for values of $T_{(i)}$, corresponding to such a block. So in Example 1.3 we immediately get the following relations, to be satisfied by the solution:

$$\hat{F}_n(T_{(1)}) = \hat{F}_n(T_{(2)}) = \hat{F}_n(T_{(3)})$$
$$\hat{F}_n(T_{(4)}) = \ldots = \hat{F}_n(T_{(7)})$$
$$\hat{F}_n(T_{(8)}) = \hat{F}_n(T_{(9)})$$
$$\hat{F}_n(T_{(10)}) = \hat{F}_n(T_{(11)}) = \hat{F}_n(T_{(12)})$$

Generally we can always take as our initial estimate a distribution function $F^{(0)}$ which is constant for values of $T_{(i)}$, corresponding blocks which have the structure described above. However, the NPMLE will typically be constant on much bigger blocks. The initial reduction gives the maximum number of blocks on which the NPMLE will be constant.

3.3 Exercises

1. Is the following statement true or false:

the EM algorithm converges, in the interval censoring problem, Case 1, to the NPMLE if and only if the starting distribution has positive mass at the points where the NPMLE puts mass.

2. Let, in the interval censoring problem, Case 1, the sample size be $n = 3$, and let $\delta_1 = \delta_2 = 1$, $\delta_3 = 0$. Take as starting distribution for the EM algorithm:

$$p_i^{(0)} = P^{(0)}\{X = T_{(i)}\} = 1/4,$$

where $T_{(1)} < T_{(2)} < T_{(3)}$ are the observation times, and $T_{(4)}$ is an arbitrary point, bigger than $T_{(3)}$. Show that the values at the nth iteration step ($n \geq 1$) are given by

$$\begin{aligned}
p_1^{(n)} &= \tfrac{2}{3} - \tfrac{1}{3} \cdot 2^{-n}, \\
p_2^{(n)} &= \tfrac{1}{3} \cdot 2^{-n}, \\
p_3^{(n)} &= 0, \\
p_4^{(n)} &= \tfrac{1}{3}.
\end{aligned}$$

3. Let, in the interval censoring problem, Case 1, $\delta_1 = 1$ and $\delta_n = 0$, and let $y_i = \hat{F}_n(T_{(i)})$, $1 \leq i \leq n$. Show that (1.9) and (1.10) follow from (3.5) and (3.6).

4. What is the analogue of (3.5) and (3.6) for interval censoring, Case 2?

5. Let the conditions of Proposition 1.4 be satisfied. Show that the distribution function $F^{(m+1)}$, maximizing (3.8), is the left derivative of the convex minorant of the cumulative sum diagram, consisting of the points

$$P_j = \left(G_{F^{(m)}}(T_{(j)}), V_{F^{(m)}}(T_{(j)}) \right),$$

where $P_0 = (0,0)$ and $T_{(j)} \in J_n$, $j = 1, 2, \ldots$.

6.(a) Describe an iterative convex minorant algorithm for the deconvolution problem, discussed in Section 2.1.

(b) What would correspond to, respectively, the E and the M step of the EM algorithm for this problem?

4 Consistency

Consistency of the NPMLE in the cases of interval censoring and deconvolution can be proved by a general method which has been used by Jewell (1982) in proving consistency of the NPMLE for the mixing distribution in scale mixtures of exponential distributions. We first illustrate the method for interval censoring, case 1.

4.1 Interval censoring, case 1

Let $(X_1, T_1), \ldots, (X_n, T_n)$ be a sample of random variables in $\mathbb{R}^2_+$, where X_i and T_i are independent (non-negative) random variables with continuous distribution functions F_0 and G, respectively, satisfying $P_{F_0} \ll P_G$ (the probability measure P_{F_0}, induced by F_0, is absolutely continuous with respect to the probability measure P_G, induced by G).

If $\hat{F}_n$ is the NPMLE, we have, for each $\epsilon \in (0, 1)$:

$$\psi\big((1 - \epsilon)\hat{F}_n + \epsilon F_0\big) - \psi(\hat{F}_n) \leq 0,$$

where $\psi(\hat{F}_n)$ is defined by (1.3). Hence we obtain

$$\lim_{\epsilon \downarrow 0} \epsilon^{-1} \left\{ \psi\big((1 - \epsilon)\hat{F}_n + \epsilon F_0\big) - \psi(\hat{F}_n) \right\} \leq 0.$$

Evaluating this limit, we obtain

$$\int \left\{ \frac{F_0(t)}{\hat{F}_n(t)} 1_{\{x \leq t\}} + \frac{1 - F_0(t)}{1 - \hat{F}_n(t)} 1_{\{x > t\}} \right\} dP_n(x, t) \leq 1. \tag{4.1}$$

We now take as our sample space Ω the space of all (infinite) sequences

$$(X_1, T_1), (X_2, T_2), \ldots,$$

and denote a point of this sample space by ω. To indicate the dependence on ω, we will write $\hat{F}_n(t; \omega)$ instead of $\hat{F}_n$, and likewise $P_n(x, t; \omega)$ instead of $P_n(x, t)$. The set Ω will be endowed with the usual Borel σ-algebra, generated by the product topology on $\prod_1^\infty \mathbb{R}^2$, and the product measure P^∞. We shall denote the latter probability measure by $\mathbb{P}$.

Fix $\epsilon \in (0, \frac{1}{2})$ and let a and b be chosen such that

$$F_0(a) = \epsilon, \quad F_0(b) = 1 - \epsilon. \tag{4.2}$$

Using the strong law of large numbers (and a separability argument), it is seen that $P_n(\cdot, \cdot; \omega)$ converges weakly to P, for all ω in a set B such that $\mathbb{P}(B) = 1$.

Fix an $\omega \in B$. By the Helly compactness theorem it is seen that the sequence of functions $\hat{F}_n(\,\cdot\,;\omega)$ has a subsequence $\hat{F}_{n_k}(\,\cdot\,;\omega)$, converging vaguely to a nondecreasing right continuous function F, taking values in $[0,1]$. Note that we cannot assume (at this point) that F is a distribution function.

We may assume that $1/\hat{F}_n(t;\omega)$ and $1/(1-\hat{F}_n(t;\omega))$ are bounded for $t \in [a,b]$ and all n sufficiently large. This follows from (4.2) and the fact that $P_n(\,\cdot\,,\cdot\,;\omega)$ converges weakly to P. Moreover, by the vague convergence of $\hat{F}_{n_k}(\,\cdot\,;\omega)$ to F we may also assume that $1/F(t)$ and $1/(1-F(t))$ are bounded for $t \in [a,b]$. Hence we assume

$$1/F(a) + 1/(1-F(b)) \leq M \tag{4.3}$$

and

$$1/\hat{F}_n(a;\omega) + 1/(1-\hat{F}_n(b;\omega)) \leq M, \tag{4.4}$$

for a constant $M > 0$ and all n sufficiently large.

We now get the following Lemma.

Lemma 4.1. Let the points a and b be chosen in such a way that (4.2) is satisfied. Then we have:

$$\lim_{k \to \infty} \int_{\mathbb{R}\times[a,b]} \left\{ \frac{F_0(t)}{\hat{F}_{n_k}(t;\omega)} 1_{\{x \leq t\}} + \frac{1-F_0(t)}{1-\hat{F}_{n_k}(t;\omega)} 1_{\{x>t\}} \right\} dP_{n_k}(x,t;\omega)$$
$$= \int_{\mathbb{R}\times[a,b]} \left\{ \frac{F_0(t)}{F(t)} 1_{\{x \leq t\}} + \frac{1-F_0(t)}{1-F(t)} 1_{\{x>t\}} \right\} dP(x,t).$$

$$\tag{4.5}$$

Moreover,

$$\int_{\mathbb{R}\times[a,b]} \left\{ \frac{F_0(t)}{F(t)} 1_{\{x \leq t\}} + \frac{1-F_0(t)}{1-F(t)} 1_{\{x>t\}} \right\} dP(x,t) \leq 1. \tag{4.6}$$

Proof. Fix $0 < \delta < 1$ and take a grid of points $a = u_0 < u_1 < \ldots < u_m = b$ on $[a,b]$ such that $m = 1 + [1/\delta^2]$ and

$$G(u_i) - G(u_{i-1}) = \{G(b) - G(a)\}/m, \; i = 1, \ldots, m.$$

First we suppose, for simplicity, that the points u_i are points of continuity for the function F, which is the (vague) limit of the sequence of functions $\hat{F}_{n_k}(\,\cdot\,;\omega)$.

Let K be the (possibly empty) set of indices i, $i = 1, \ldots, m$ such that

$$\max\{1/F(u_{i-1}) - 1/F(u_i), 1/(1-F(u_i)) - 1/(1-F(u_{i-1}))\} \geq \delta.$$

By (4.3), the number of indices of this type is not bigger than $1 + [M/\delta]$. Let L be the remaining set of indices i, $i = 1, \ldots, m$. Denoting the interval $[u_0, u_1]$ by J_1

and the intervals $(u_{i-1}, u_i]$ by J_i, $i > 1$, we get

$$\int_{\mathbb{R}\times[a,b]} \left\{ \frac{F_0(t)}{\hat{F}_{n_k}(t\,;\,\omega)} 1_{\{x\leq t\}} + \frac{1 - F_0(t)}{1 - \hat{F}_{n_k}(t\,;\,\omega)} 1_{\{x>t\}} \right\} dP_{n_k}(x,t\,;\,\omega)$$
$$= \sum_{i=1}^{m} \int_{\mathbb{R}\times J_i} \left\{ \frac{F_0(t)}{\hat{F}_{n_k}(t\,;\,\omega)} 1_{\{x\leq t\}} + \frac{1 - F_0(t)}{1 - \hat{F}_{n_k}(t\,;\,\omega)} 1_{\{x>t\}} \right\} dP_{n_k}(x,t\,;\,\omega).$$

Since $\hat{F}_{n_k}(u_i\,;\,\omega)$ converges to $F(u_i)$ for each i, $0 \leq i \leq m$, we get, for sufficiently large k:

$$\begin{aligned}
1/\hat{F}_{n_k}(u_{i-1}\,;\,\omega) - 1/\hat{F}_{n_k}(u_i\,;\,\omega) &< 2\delta, \ i \in L \\
1/(1 - \hat{F}_{n_k}(u_i\,;\,\omega)) - 1/(1 - \hat{F}_{n_k}(u_{i-1}\,;\,\omega)) &< 2\delta, \ i \in L.
\end{aligned} \tag{4.7}$$

Hence

$$\int_{\mathbb{R}\times[a,b]} \left\{ \frac{F_0(t)}{\hat{F}_{n_k}(t\,;\,\omega)} 1_{\{x\leq t\}} + \frac{1 - F_0(t)}{1 - \hat{F}_{n_k}(t\,;\,\omega)} 1_{\{x>t\}} \right\} dP_{n_k}(x,t\,;\,\omega)$$
$$= \sum_{i\in K} \int_{\mathbb{R}\times J_i} \left\{ \frac{F_0(t)}{\hat{F}_{n_k}(t\,;\,\omega)} 1_{\{x\leq t\}} + \frac{1 - F_0(t)}{1 - \hat{F}_{n_k}(t\,;\,\omega)} 1_{\{x>t\}} \right\} dP_{n_k}(x,t\,;\,\omega)$$
$$+ \sum_{i\in L} \int_{\mathbb{R}\times J_i} \left\{ \frac{F_0(t)}{\hat{F}_{n_k}(t\,;\,\omega)} 1_{\{x\leq t\}} + \frac{1 - F_0(t)}{1 - \hat{F}_{n_k}(t\,;\,\omega)} 1_{\{x>t\}} \right\} dP_{n_k}(x,t\,;\,\omega)$$
$$= \int_{\mathbb{R}\times[a,b]} \left\{ \frac{F_0(t)}{\hat{F}_{n_k}(t\,;\,\omega)} 1_{\{x\leq t\}} + \frac{1 - F_0(t)}{1 - \hat{F}_{n_k}(t\,;\,\omega)} 1_{\{x>t\}} \right\} dP(x,t) + r_k(\omega),$$
$$\tag{4.8}$$

where $|r_k(\omega)| \leq c\cdot\delta$, for a constant $c > 0$. This can be seen by replacing $\hat{F}_{n_k}(t\,;\,\omega)$ on each interval J_i by its value $\hat{F}_{n_k}(u_i\,;\,\omega)$ at the right endpoint of the interval, and by noting that for large k

$$\left| 1/\hat{F}_{n_k}(t\,;\,\omega) - 1/\hat{F}_{n_k}(u_i\,;\,\omega) \right| < 2\delta,$$

if $i \in L$ (with a similar inequality for $1/\{1 - \hat{F}_{n_k}(t\,;\,\omega)\}$). On the intervals J_i, with $i \in K$, we use (4.4). Note that

$$\sum_{i\in K} P(\mathbb{R} \times J_i) \to 0, \ \text{if } \delta \downarrow 0,$$

since $P(\mathbb{R} \times J_i)$ is of order $\mathcal{O}(\delta^2)$, while the number of intervals J_i such that $i \in K$ is of order $\mathcal{O}(1/\delta)$.

On the other hand we have by dominated convergence:

$$\lim_{k\to\infty} \int_{\mathbb{R}\times[a,b]} \left\{ \frac{F_0(t)}{\hat{F}_{n_k}(t\,;\,\omega)} 1_{\{x\leq t\}} + \frac{1 - F_0(t)}{1 - \hat{F}_{n_k}(t\,;\,\omega)} 1_{\{x>t\}} \right\} dP(x,t)$$
$$= \int_{\mathbb{R}\times[a,b]} \left\{ \frac{F_0(t)}{F(t)} 1_{\{x\leq t\}} + \frac{1 - F_0(t)}{1 - F(t)} 1_{\{x>t\}} \right\} dP(x,t).$$
$$\tag{4.9}$$

Combining (4.8) and (4.9) we obtain

$$\int_{\mathbb{R}\times[a,b]}\left\{\frac{F_0(t)}{\hat{F}_{n_k}(t;\omega)}1_{\{x\le t\}}+\frac{1-F_0(t)}{1-\hat{F}_{n_k}(t;\omega)}1_{\{x>t\}}\right\}dP_{n_k}(x,t;\omega)$$
$$=\int_{\mathbb{R}\times[a,b]}\left\{\frac{F_0(t)}{F(t)}1_{\{x\le t\}}+\frac{1-F_0(t)}{1-F(t)}1_{\{x>t\}}\right\}dP(x,t)+r_k'(\omega),\qquad(4.10)$$

where $\left|r_k'(\omega)\right|\le c'\delta$.

If one or more of the points u_i is not a point of continuity of F, we shift the point u_i a bit to the left or right, in order to get continuity points (using the fact that the continuity points of F are dense). So in all cases we get a relation of type (4.10). Since δ can be chosen arbitrarily small, (4.5) now follows. Relation (4.6) immediately follows from (4.5) and (4.1).   □

By monotone convergence we now obtain from (4.6):

$$\int_{\mathbb{R}^2}\left\{\frac{F_0(t)}{F(t)}1_{\{x\le t\}}+\frac{1-F_0(t)}{1-F(t)}1_{\{x>t\}}\right\}dP(x,t)$$
$$=\lim_{a\downarrow0,b\to\infty}\int_{\mathbb{R}\times[a,b]}\left\{\frac{F_0(t)}{F(t)}1_{\{x\le t\}}+\frac{1-F_0(t)}{1-F(t)}1_{\{x>t\}}\right\}dP(x,t)\le 1.\qquad(4.11)$$

This, however, can only happen if $F=F_0$. For we can write

$$\int_{\mathbb{R}^2}\left\{\frac{F_0(t)}{F(t)}1_{\{x\le t\}}+\frac{1-F_0(t)}{1-F(t)}1_{\{x>t\}}\right\}dP(x,t)$$
$$=\int_{\mathbb{R}}\left\{\frac{F_0(t)^2}{F(t)}+\frac{(1-F_0(t))^2}{1-F(t)}\right\}dG(t),$$

and the latter expression is strictly bigger than 1, unless $F=F_0$.

This is proved in the following way. If $0<F_0(t)<1$ and $y\in(0,1)$, then

$$\frac{F_0(t)^2}{y}+\frac{(1-F_0(t))^2}{1-y}=\begin{cases}1,&\text{if }y=F_0(t)\\>1,&\text{if }y\ne F_0(t).\end{cases}\qquad(4.12)$$

By the monotonicity of F, the monotonicity and continuity of F_0, and the absolute continuity of P_{F_0} with respect to P_G, we have

$$F\ne F_0\Rightarrow F(t)\ne F_0(t)\text{ on an interval of increase of }G.$$

Hence by (4.12)

$$\int_{\mathbb{R}}\left\{\frac{F_0(t)^2}{F(t)}+\frac{(1-F_0(t))^2}{1-F(t)}\right\}dG(t)>1,$$

if $F\ne F_0$. This contradicts (4.11).

Thus we have proved that for each ω outside a set of probability zero, each subsequence of the sequence of functions $\hat{F}_n(\cdot\,;\omega)$ has a vaguely convergent sequence, and that all these convergent subsequences have the same limit F_0. This proves that the sequence $\hat{F}_n$ converges weakly to F_0, with probability one. Since F_0 is continuous, this is the same as saying that $\hat{F}_n$ converges with probability one to F_0 in the supremum distance on the set of distribution functions, i.e.

$$\mathbb{P}\left\{\lim_{n\to\infty}\sup_{t\in\mathbb{R}}\left|\hat{F}_n(t)-F_0(t)\right|=0\right\}=1.$$

We next illustrate the method for the deconvolution problem.

4.2 Convolution with a symmetric density

Let $Z_1,\ldots,Z_n$ be a sample from a distribution with density

$$z\mapsto\int g(z-y)\,dF_0(y),\ z\in\mathbb{R},$$

where g is a symmetric density, satisfying (2.14) and (2.15), and where F_0 is an unknown continuous distribution function.

Proceeding as before, we find

$$\lim_{\epsilon\downarrow0}\epsilon^{-1}\left\{\psi\big((1-\epsilon)\hat{F}_n+\epsilon F_0\big)-\psi(\hat{F}_n)\right\}\le0,$$

implying

$$\int\frac{\int g(z-y)\,dF_0(y)}{\int g(z-y)\,d\hat{F}_n(y)}\,dH_n(z)\le1. \tag{4.13}$$

For each ω in a set B of probability 1, the sequence $H_n(\cdot\,;\omega)$ converges weakly to the distribution function H_0, with density

$$h_0(z)=\int g(z-y)\,dF_0(y),\ z\in\mathbb{R}.$$

Since H_0 is continuous, we have

$$\sup_{z\in\mathbb{R}}\left|H_n(z;\omega)-H_0(z)\right|\to0,\text{ as }n\to\infty.$$

Fix $\omega\in B$ and consider a nonempty closed set of the form

$$A_\epsilon=\left\{z\in\mathbb{R}:\int g(z-y)\,dF_0(y)\ge\epsilon\right\},$$

where $\epsilon > 0$. The sequence of functions $\hat{F}_n(\,\cdot\,;\omega)$ has a subsequence $\left(\hat{F}_{n_k}(\,\cdot\,;\omega)\right)_{k=1}^{\infty}$, converging vaguely to a subdistribution function F, and there exists a $\delta > 0$ such that

$$\inf_{z \in A_\epsilon} \int g(z - y)\, d\hat{F}_{n_k}(y\,;\omega) \geq \delta, \tag{4.14}$$

for each sufficiently large k, since otherwise (4.13) would be violated. We get the following lemma, analogous to Lemma 4.1.

Lemma 4.2. We have:

$$\lim_{k \to \infty} \int_{A_\epsilon} \frac{\int g(z - y)\, dF_0(y)}{\int g(z - y)\, d\hat{F}_{n_k}(y\,;\omega)}\, dH_{n_k}(z\,;\omega)$$

$$= \int_{A_\epsilon} \frac{\int g(z - y)\, dF_0(y)}{\int g(z - y)\, dF(y)}\, dH_0(z). \tag{4.15}$$

Moreover,

$$\int_{A_\epsilon} \frac{\int g(z - y)\, dF_0(y)}{\int g(z - y)\, dF(y)}\, dH_0(z) \leq 1. \tag{4.16}$$

Proof. Since $\hat{F}_{n_k}(\,\cdot\,;\omega)$ converges vaguely to F, we have

$$\lim_{k \to \infty} \int g(z - y)\, d\hat{F}_{n_k}(y\,;\omega) = \int g(z - y)\, dF(y),$$

Define the functions h_n and h by

$$h_n(z) = \int g(z - y)\, d\hat{F}_n(y\,;\omega), \; h(z) = \int g(z - y)\, dF(y), \; z \in \mathbb{R}.$$

Since g is bounded and uniformly continuous on $\mathbb{R}$, the sequence of continuous functions h_{n_k} converges uniformly to the (continuous) limit function h. Moreover, by (4.14)

$$\inf_{z \in A_\epsilon} h(z) \geq \delta > 0.$$

(note that A_ϵ is a compact set).

It now follows that the functions

$$z \mapsto \int g(z - y)\, dF_0(y)/h_{n_k}(z), \; z \in A_\epsilon$$

are bounded continuous functions, converging uniformly to the bounded continuous function

$$z \mapsto \int g(z - y)\, dF_0(y)/h(z), \; z \in A_\epsilon.$$

Since $H_{n_k}(\,\cdot\,;\omega)$ converges weakly to H_0, (4.15) easily follows, and (4.16) is implied by (4.15).   □

We now proceed as before. The monotone convergence theorem and (4.16) imply

$$\int \frac{\int g(z-y)\,dF_0(y)}{\int g(z-y)\,dF(y)}\,dH_0(z) \leq 1. \qquad (4.17)$$

This can only happen if $F = F_0$. For we get from (4.17):

$$\int h_0(z)^2/h(z)\,dz \leq 1,$$

where h_0 is the density of H_0. But $\int h_0(z)^2/h(z)\,dz$ is minimized by taking $h = h_0$. This is seen by first observing that we may assume that

$$\int h(z)\,dz = 1,$$

since otherwise we could make the integral $\int h_0(z)^2/h(z)\,dz$ smaller by multiplying h by a constant bigger than 1, and secondly by observing that the integrand of the integral

$$\int \left\{ \frac{h_0(z)^2}{h(z)} + h(z) \right\}\,dz$$

is pointwise minimized by taking $h(z) = h_0(z)$ for each z.

Now $F \neq F_0$ would imply

$$h(z) = \int g(z-y)\,dF(y) \neq h_0(z),$$

for z in an interval of positive length, and hence

$$\int \frac{\int g(z-y)\,dF_0(y)}{\int g(z-y)\,dF(y)}\,dH_0(z) = \int \frac{h_0(z)^2}{h(z)}\,dz > 1, \qquad (4.18)$$

contradicting (4.17).

We can conclude from this that every subsequence of the sequence of functions $\hat{F}_n(\,\cdot\,;\omega)$ has a convergent subsequence, and that all these subsequences have the same (weak) limit F_0. This implies the consistency of the NPMLE.

We finally demonstrate the method for the more complicated case of interval censoring.

4.3 Interval censoring, case 2

Let $(X_1, T_1, U_1), \ldots, (X_n, T_n, U_n)$ be a sample of random variables in $\mathbb{R}_+^3$, where X_i is a (non-negative) random variable with continuous df F_0, and where T_i and U_i

are (non-negative) random variables, independent of X_i, with a joint continuous distribution function H and such that $T_i \leq U_i$ with probability one. Moreover, we asume that H has a density h with respect to Lebesgue measure, satisfying

$$h(t, u) > 0, \text{ if } 0 < F_0(t) < F_0(u) < 1. \tag{4.19}$$

As in section 4.1, we obtain

$$\lim_{\epsilon \downarrow 0} \epsilon^{-1} \left\{ \psi\big((1 - \epsilon)\hat{F}_n + \epsilon F_0\big) - \psi(\hat{F}_n) \right\} \leq 0,$$

implying

$$\int \left\{ \frac{F_0(t)}{\hat{F}_n(t)} 1_{\{x \leq t\}} + \frac{F_0(u) - F_0(t)}{\hat{F}_n(u) - \hat{F}_n(t)} 1_{\{t < x \leq u\}} + \frac{1 - F_0(u)}{1 - \hat{F}_n(u)} 1_{\{x > u\}} \right\} dP_n(x, t, u) \leq 1. \tag{4.20}$$

We now proceed as in section 4.1. Using the strong law of large numbers, it is seen that $P_n(\cdot, \cdot, \cdot; \omega)$ converges weakly to P, for all ω in a set B such that $\mathbb{P}(B) = 1$. Fix an $\omega \in B$. By the Helly compactness theorem it is seen that the sequence of functions $\hat{F}_n(\cdot; \omega)$ has a subsequence $\hat{F}_{n_k}(\cdot; \omega)$, converging vaguely to a subdistribution function F.

Fix $\epsilon \in (0, 1/2)$ and define the set A_ϵ by

$$A_\epsilon = \{(t, u) : F_0(t) \geq \epsilon, F_0(u) - F_0(t) \geq \epsilon, 1 - F_0(u) \geq \epsilon\}.$$

Moreover, let a and b be chosen such that

$$F_0(a) = \epsilon, F_0(b) = 1 - \epsilon. \tag{4.21}$$

By (4.20) we may assume that there exists an $M > 0$ such that

$$1/\hat{F}_n(t; \omega) + 1/\big(\hat{F}_n(u; \omega) - \hat{F}_n(t; \omega)\big) + 1/\big(1 - \hat{F}_n(u; \omega)\big) \leq M, \tag{4.22}$$

for $(t, u) \in A_\epsilon$ and for all sufficiently large n. Note that the left side of (4.22) cannot tend to ∞ at points $(t, u) \in A_\epsilon$ such that $F_0(u) - F_0(t) = \epsilon$, since in that case the left side of (4.22) would become arbitrarily large on a subset of $A_{\epsilon/2}$ with Lebesgue measure bounded away from zero, as $n \to \infty$, which cannot happen because of (4.19), (4.21), the continuity of F_0 and the weak convergence of $P_n(\cdot, \cdot, \cdot; \omega)$ to P.

We get the following lemma, analogous to Lemma 4.1.

Lemma 4.3. We have:

$$\begin{aligned}
\lim_{k \to \infty} &\int_{\mathbb{R} \times A_\epsilon} \left\{ \frac{F_0(t)}{\hat{F}_{n_k}(t; \omega)} 1_{\{x \leq t\}} + \frac{F_0(u) - F_0(t)}{\hat{F}_{n_k}(u; \omega) - \hat{F}_{n_k}(t; \omega)} 1_{\{t < x \leq u\}} \right. \\
&\left. + \frac{1 - F_0(u)}{1 - \hat{F}_{n_k}(u; \omega)} 1_{\{x > u\}} \right\} dP_{n_k}(x, t, u; \omega) \\
= &\int_{\mathbb{R} \times A_\epsilon} \left\{ \frac{F_0(t)}{F(t)} 1_{\{x \leq t\}} + \frac{F_0(u) - F_0(t)}{F(u) - F(t)} 1_{\{t < x \leq u\}} \right. \\
&\left. + \frac{1 - F_0(u)}{1 - F(u)} 1_{\{x > u\}} \right\} dP(x, t, u).
\end{aligned} \tag{4.23}$$

Moreover,

$$\int_{\mathbb{R} \times A_\epsilon} \left\{ \frac{F_0(t)}{F(t)} 1_{\{x \le t\}} + \frac{F_0(u) - F_0(t)}{F(u) - F(t)} 1_{\{t < x \le u\}} \right.$$
$$\left. + \frac{1 - F_0(u)}{1 - F(u)} 1_{\{x > u\}} \right\} dP(x, t, u) \le 1. \tag{4.24}$$

Proof. Let the points a and b be defined by (4.21). We fix $0 < \delta < 1$ and consider a grid of points t_i, $t_0 < \ldots < t_m$, such that $m = 1 + \left[1/\delta^2 \right]$, $t_0 = a$, $t_m = b$, and

$$\int_{t \in [t_{i-1}, t_i]} h(t, u) \, dt \, du = m^{-1} \int_{t \in [a, b]} h(t, u) \, dt, \ i = 1, \ldots, m.$$

Likewise, we define a grid of points u_i, $u_0 < \ldots < u_m$, such that $u_0 = a$, $u_m = b$, and

$$\int_{u \in [u_{i-1}, u_i]} h(t, u) \, dt \, du = m^{-1} \int_{u \in [a, b]} h(t, u) \, dt, \ i = 1, \ldots, m.$$

We denote the intervals $(t_{i-1}, t_i]$ by J_i and the intervals $(u_{i-1}, u_i]$ by J_i'. Furthermore, let K be the (possibly empty) set of indices i, $i = 1, \ldots, m$ such that

$$\max \left\{ 1/F(t_{i-1}) - 1/F(t_i), 1/(1 - F(t_i)) - 1/(1 - F(t_{i-1})) \right\} \ge \delta,$$

and let L be the remaining set of indices i of the points t_i. Likewise, let K' be the (possibly empty) set of indices i, $i = 1, \ldots, m$ such that

$$\max \left\{ 1/F(u_{i-1}) - 1/F(u_i), 1/(1 - F(u_i)) - 1/(1 - F(u_{i-1})) \right\} \ge \delta,$$

and let L' be the remaining set of indices i of the points u_i.

First suppose that the t_i and u_i are continuity points of F. The term

$$\int_{\mathbb{R} \times A_\epsilon} \left\{ \frac{F_0(t)}{\hat{F}_{n_k}(t; \omega)} 1_{\{x \le t\}} + \frac{1 - F_0(u)}{1 - \hat{F}_{n_k}(u; \omega)} 1_{\{x > u\}} \right\} dP_{n_k}(x, t, u; \omega)$$

can be treated as before, showing that

$$\int_{\mathbb{R} \times A_\epsilon} \left\{ \frac{F_0(t)}{\hat{F}_{n_k}(t; \omega)} 1_{\{x \le t\}} + \frac{1 - F_0(u)}{1 - \hat{F}_{n_k}(u; \omega)} 1_{\{x > u\}} \right\} dP_{n_k}(x, t, u; \omega)$$
$$= \int_{\mathbb{R} \times A_\epsilon} \left\{ \frac{F_0(t)}{\hat{F}_{n_k}(t; \omega)} 1_{\{x \le t\}} + \frac{1 - F_0(u)}{1 - \hat{F}_{n_k}(u; \omega)} 1_{\{x > u\}} \right\} dP(x, t, u) + r_k(\omega),$$

where $|r_k(\omega)| \le c \cdot \delta$, for a constant $c > 0$. Similarly, for the "middle term"

$$\int_{\mathbb{R} \times A_\epsilon} \left\{ \frac{F_0(u) - F_0(t)}{\hat{F}_{n_k}(u; \omega) - \hat{F}_{n_k}(t; \omega)} 1_{\{t < x \le u\}} \right\} dP_{n_k}(x, t, u; \omega)$$
$$= \int_{\mathbb{R} \times A_\epsilon} \left\{ \frac{F_0(u) - F_0(t)}{\hat{F}_{n_k}(u; \omega) - \hat{F}_{n_k}(t; \omega)} 1_{\{t < x \le u\}} \right\} dP(x, t, u) + r_k'(\omega),$$

where $|r'_k(\omega)| \leq c \cdot \delta$, for a constant $c > 0$. This is seen by replacing $\hat{F}_{n_k}(t\,;\omega)$ on each interval J_i by its value $\hat{F}_{n_k}(t_{i-1}\,;\omega)$ at the *left* endpoint of the interval, and by replacing $\hat{F}_{n_k}(u\,;\omega)$ on each interval J'_i by its value $\hat{F}_{n_k}(u_i\,;\omega)$ at the *right* endpoint of the interval, and noting that for large k, and $(t,u) \in A_\epsilon$:

$$\left|1/\{\hat{F}_{n_k}(u\,;\omega) - \hat{F}_{n_k}(t_{i-1}\,;\omega)\} - 1/\{\hat{F}_{n_k}(u\,;\omega) - \hat{F}_{n_k}(t\,;\omega)\}\right| < 2\delta M^2,$$

if $i \in L$ and that

$$\left|1/\{\hat{F}_{n_k}(u\,;\omega) - \hat{F}_{n_k}(t\,;\omega)\} - 1/\{\hat{F}_{n_k}(u_i\,;\omega) - \hat{F}_{n_k}(t\,;\omega)\}\right| < 2\delta M^2,$$

if $i \in L'$.

If $t \in J_i$, $i \in K$ or $u \in J'_i$, $i \in K'$, we use (4.22).

Note that

$$\sum_{i \in K} \int_{\mathbb{R} \times J_i \times \mathbb{R}} dP(x,t,u) \to 0, \text{ if } \delta \downarrow 0,$$

and likewise

$$\sum_{i \in K'} \int_{\mathbb{R}^2 \times J'_i} dP(x,t,u) \to 0, \text{ if } \delta \downarrow 0,$$

since $P(\mathbb{R} \times J_i \times \mathbb{R})$ and $P(\mathbb{R}^2 \times J'_i)$ are of order $\mathcal{O}(\delta^2)$, while the number of intervals J_i such that $i \in K$ is of order $\mathcal{O}(1/\delta)$, and likewise the number of intervals J'_i such that $i \in K'$ is of order $\mathcal{O}(1/\delta)$.

By dominated convergence we get

$$\lim_{k \to \infty} \int_{\mathbb{R} \times A_\epsilon} \left\{ \frac{F_0(t)}{\hat{F}_{n_k}(t\,;\omega)} 1_{\{x \leq t\}} + \frac{F_0(u) - F_0(t)}{\hat{F}_{n_k}(u\,;\omega) - \hat{F}_{n_k}(t\,;\omega)} 1_{\{t < x \leq u\}} \right.$$
$$\left. + \frac{1 - F_0(u)}{1 - \hat{F}_{n_k}(u\,;\omega)} 1_{\{x > u\}} \right\} dP(x,t,u\,;\omega)$$
$$\tag{4.25}$$
$$= \int_{\mathbb{R} \times A_\epsilon} \left\{ \frac{F_0(t)}{F(t)} 1_{\{x \leq t\}} + \frac{F_0(u) - F_0(t)}{F(u) - F(t)} 1_{\{t < x \leq u\}} \right.$$
$$\left. + \frac{1 - F_0(u)}{1 - F(u)} 1_{\{x > u\}} \right\} dP(x,t,u).$$

If one or more of the points of the (two) grids is not a point of continuity of F, we argue as in the proof of Lemma 4.1. This yields (4.23), and (4.24) easily follows. $\square$

By monotone convergence we now obtain from (4.24):

$$
\int_{\mathbb{R}^3} \left\{ \frac{F_0(t)}{F(t)} 1_{\{x \leq t\}} + \frac{F_0(u) - F_0(t)}{F(u) - F(t)} 1_{\{t < x \leq u\}} \right.
$$
$$
\left. + \frac{1 - F_0(u)}{1 - F(u)} 1_{\{x > u\}} \right\} dP(x, t, u)
$$
$$
= \lim_{\epsilon \downarrow 0} \int_{\mathbb{R} \times A_\epsilon} \left\{ \frac{F_0(t)}{F(t)} 1_{\{x \leq t\}} + \frac{F_0(u) - F_0(t)}{F(u) - F(t)} 1_{\{t < x \leq u\}} \right. \tag{4.26}
$$
$$
\left. + \frac{1 - F_0(u)}{1 - F(u)} 1_{\{x > u\}} \right\} dP(x, t, u)
$$
$$
\leq 1.
$$

This, however, can again only happen if $F = F_0$. For we can write

$$
\int_{\mathbb{R}^3} \left\{ \frac{F_0(t)}{F(t)} 1_{\{x \leq t\}} + \frac{F_0(u) - F_0(t)}{F(u) - F(t)} 1_{\{t < x \leq u\}} + \frac{1 - F_0(u)}{1 - F(u)} 1_{\{x > u\}} \right\} dP(x, t, u)
$$
$$
= \int_{\mathbb{R}^2} \left\{ \frac{F_0(t)^2}{F(t)} + \frac{(F_0(u) - F_0(t))^2}{F(u) - F(t)} + \frac{(1 - F_0(u))^2}{1 - F(u)} \right\} dH(t, u),
$$

and the latter expression is strictly bigger than 1, unless $F = F_0$.

The proof is analogous to the proof for interval censoring, case 1. If $0 < F_0(t) < F_0(u) < 1$ and $0 < x < y < 1$, then

$$
\frac{F_0(t)^2}{x} + \frac{(F_0(u) - F_0(t))^2}{y - x} + \frac{(1 - F_0(u))^2}{1 - y} = \begin{cases} 1, & \text{if } x = F_0(t), \, y = F_0(u) \\ > 1, & \text{otherwise} \end{cases}
$$
$$
\tag{4.27}
$$

By the monotonicity of F, the monotonicity and continuity of F_0, and (4.19) we now have

$$
\int_{\mathbb{R}^2} \left\{ \frac{F_0(t)^2}{F(t)} + \frac{(F_0(u) - F_0(t))^2}{F(u) - F(t)} + \frac{(1 - F_0(u))^2}{1 - F(u)} \right\} dH(t, u) > 1,
$$

if $F \neq F_0$. This contradicts (4.26).

As in the case of interval censoring, case 1, we get

$$
P \left\{ \lim_{n \to \infty} \sup_{t \in \mathbb{R}} \left| \hat{F}_n(t) - F_0(t) \right| = 0 \right\} = 1.
$$

4.4 Exercises

1. Let F_0 be a continuous distribution function of a non-negative random variable, and let g be a decreasing probability density on $[0, \infty)$, which is continuous on $[0, \infty)$. Show by the method of this chapter that the NPMLE is strongly consistent for the deconvolution model, consisered in section 2.1.

The following exercises give an idea of the "entropy approach" in proving consistency. We consider the interval censoring model, case 1, under the continuity assumptions, specified in section 4.1.

2. Let $\mathcal{F}$ be the set of distribution functions on $\mathbb{R}$, and let $\mathcal{A}$ be the class of functions $\{\phi_F : F \in \mathcal{F}\}$, where

$$\phi_F(x, t) = \{x \le t\} \sqrt{F(t)} + \{x > t\} \sqrt{1 - F(t)} \quad , x, t \in \mathbb{R}.$$

Let $N_2(\delta, P_n, \mathcal{A})$ be the minimum number of balls with radius $\delta > 0$, needed to cover the set $\mathcal{A}$, using the L_2-distance

$$\|\phi_F - \phi_G\|_2 \stackrel{\text{def}}{=} \left\{ \int |\phi_F(x, t) - \phi_G(x, t)|^2 \, dP_n(x, t) \right\}^{1/2}, \ F, G \in \mathcal{F},$$

where P_n is the empirical measure of $(X_1, T_1), \ldots, (X_n, T_n)$. Show that

$$\mathbb{P}\left\{ \lim_{n \to \infty} n^{-1} \log N_2(\delta, P_n, \mathcal{A}) = \prime \right\} = \infty.$$

3. Let $\mathcal{A}_\prime$ be the class of functions $\psi_F : F \in \mathcal{F}\}$, where

$$\psi_F(x, t) = 1_{\{(x,t):\phi_{F_0}(x,t)>0\}} \phi_F(x, t)/\phi_{F_0}(x, t) \quad , x, t \in \mathbb{R},$$

and let $N_1(\delta, P_n, \mathcal{A})$ be the minimum number of balls with radius $\delta > 0$, needed to cover the set $\mathcal{A}$, using the L_1-distance

$$\|\phi_F - \phi_G\|_1 \stackrel{\text{def}}{=} \int |\phi_F(x, t) - \phi_G(x, t)| \, dP_n(x, t), \ F, G \in \mathcal{F}.$$

Show

$$\mathbb{P}\left\{ \lim_{n \to \infty} n^{-1} \log N_1(\delta, P_n, \mathcal{A}) = \prime \right\} = \infty.$$

(**Hint:** use Cauchy-Schwarz.)

4. Deduce from Exercise 3:

$$\sup_{F \in \mathcal{F}} \int \psi_F(x, t) \, d(P_n - P)(x, t) \to 0,$$

with probability one.

5. Let $\hat{F}_n$ be the NPMLE of F_0. Show

$$0 \leq \int \log(\psi^2_{\hat{F}_n}) \, dP_n \leq 2 \int \psi_{\hat{F}_n} \, dP_n - 2$$
$$= 2 \int \psi_{\hat{F}_n} \, d(P_n - P)(x,t) - 2 \int \left\{ \psi_{\hat{F}_n} - \psi_{F_0} \right\}^2 \, dP. \tag{4.28}$$

Deduce from Exercise 4 and (4.28) that, with probability one:

$$\int \left\{ \psi_{\hat{F}_n} - \psi_{F_0} \right\}^2 \, dP \to 0, \text{ as } n \to \infty. \tag{4.29}$$

6. Derive from (4.29) the strong consistency of $\hat{F}_n$. Can you give an interpretation of the left-hand side of (4.29) in terms of a (squared) Hellinger distance?

Remark. The outline of the "entropy proof" of the consistency of the NPMLE in the case of interval censoring, case 1, is based on a personal communication by Sara van de Geer. For extensions and other approaches, see van de Geer (1990); in the latter report also rates of convergence are derived by using entropy methods.

5 Distribution Theory

Parts of the present chapter have a somewhat heuristic character in the sense that we use a working hypothesis which can be formulated as follows.

Working hypothesis. Starting with the real underlying distribution function F_0, the iterative convex minorant algorithm will give at the first iteration step an estimator which is asymptotically equivalent with the maximum likelihood estimator.

By "asymptotically equivalent" we mean the following. Suppose $F_n^{(1)}$ is the estimator of F_0, obtained at the first step of the iterative convex minorant algorithm, with starting distribution $F^{(0)} = F_0$, and suppose

$$\alpha_n\big(F_n^{(1)}(t) - F_0(t)\big) \xrightarrow{\mathcal{D}} Z, \text{ as } n \to \infty,$$

where Z is some nondegenerate random variable, t is a fixed point in the interior of the support of F_0, α_n is a norming constant, and where $\xrightarrow{\mathcal{D}}$ denotes convergence in distribution. Then also

$$\alpha_n\big(\hat{F}_n(t) - F_0(t)\big) \xrightarrow{\mathcal{D}} Z, \text{ as } n \to \infty,$$

where $\hat{F}_n$ is the maximum likelihood estimator of F_0.

The working hypothesis is supported by computer experiments, and certainly holds (under some condition on F_0 and G in the neighborhood of t) for interval censoring, Case 1, but so far there is no proof for the general situation. Roughly speaking, the fact one would have to show is that "off-diagonal elements of the matrix of second derivatives of the log likelihood function can be neglected" (some arguments for why this might be true will be given below).

We will first demonstrate the method for interval censoring, Case 1. In this situation the off-diagonal elements of the matrix of second derivatives of the log likelihood function are zero. So we don't have to deal with the problem mentioned in the preceding paragraph, and still can demonstrate the general method.

5.1 Interval censoring, Case 1

We will use the same set-up as in Section 4.1. The following result will be proved.

Theorem 5.1. Let t_0 be such that $0 < F_0(t_0), G(t_0) < 1$, and let F_0 and G be differentiable at t_0, with strictly positive derivatives $f_0(t_0)$ and $g(t_0)$, respectively. Furthermore, let $\hat{F}_n$ be the NPMLE of F_0. Then we have, as $n \to \infty$,

$$n^{1/3}\{\hat{F}_n(t_0) - F_0(t_0)\}/\big\{\tfrac{1}{2}F_0(t_0)(1 - F_0(t_0))f_0(t_0)/g(t_0)\big\}^{1/3} \xrightarrow{\mathcal{D}} 2Z,$$

where $\xrightarrow{\mathcal{D}}$ denotes convergence in distribution, and where Z is the last time where standard two-sided Brownian motion minus the parabola $y(t) = t^2$ reaches its maximum.

Remark 5.1. A straightforward proof of theorem 5.1 is outlined in exercises 1 to 4 at the end of this chapter. The method used there does not lead to suggestions for distribution theory for interval censoring, case 2, in contrast with the method given below. For this reason we use a method that may seem a little bit of a detour (compared to the method of exercises 1 to 4). Still another proof is given in Groeneboom (1987).

We first study the limiting behavior of $F_n^{(1)}$. To this end, we consider the process $W_n^{(0)}$, defined by

$$W_n^{(0)}(t) = \int_{t' \in [0,t]} \left\{ F_0(t')^{-1} 1_{\{x \leq t'\}} - \left(1 - F_0(t')\right)^{-1} 1_{\{x > t'\}} \right\} dP_n(x, t'), \ t \geq 0,$$

$$(5.1)$$

and the process $G_n^{(0)}$, defined by

$$G_n^{(0)}(t) = \int_{t' \in [0,t]} \left\{ F_0(t')^{-2} 1_{\{x \leq t'\}} + \left(1 - F_0(t')\right)^{-2} 1_{\{x > t'\}} \right\} dP_n(x, t'), \ t \geq 0,$$

$$(5.2)$$

i.e., $W_n^{(0)}$ and $G_n^{(0)}$ are defined in a similar way as the processes W_F and G_F, for interval censoring, Case 2 (see (1.25) and (1.29)), with $F = F_0$.

Furthermore, we define the process $V_n^{(0)}$ by

$$V_n^{(0)}(t) = W_n^{(0)}(t) + \int_{[0,t]} \left(F_0(t') - F_0(t_0)\right) dG_n^{(0)}(t'), \ t \geq 0. \qquad (5.3)$$

The process $V_n^{(0)}$ has the following property.

Lemma 5.1. Let the process $U_n^{(0)}$ be defined by

$$U_n^{(0)}(t) = n^{2/3} \left\{ V_n^{(0)}(t_0 + n^{-1/3}t) - V_n^{(0)}(t_0) \right\}, \ t \in \mathbb{R},$$

where $U_n^{(0)}(t) = 0$, if $t \leq -t_0 n^{1/3}$. Then $U_n^{(0)}$ converges in distribution, in the topology of uniform convergence on compacta on the space of locally bounded real-valued functions on $\mathbb{R}$, to the process U, defined by

$$U(t) = \sqrt{g(t_0) / \left(F_0(t_0)(1 - F_0(t_0))\right)} \, W(t) + \frac{1}{2} \frac{f_0(t_0) g(t_0)}{F_0(t_0)\left(1 - F_0(t_0)\right)} t^2, \ t \in \mathbb{R}, \ (5.4)$$

where W is (standard) two-sided Brownian motion on $\mathbb{R}$, originating from zero.

Proof. For $t \geq 0$ we can write:

$$U_n^{(0)}(t) = P_n g_n(\cdot, \cdot; t), \qquad (5.5)$$

where

$$g_n(x,u\,;t) = n^{2/3} 1_{(t_0,t_0+n^{-1/3}t]}(u)\left\{ \frac{\{x \le u\}}{F_0(u)} - \frac{\{x > u\}}{1 - F_0(u)} \right.$$
$$\left. + \frac{\{x \le u\}\big(F_0(u) - F_0(t_0)\big)}{F_0(u)^2} + \frac{\{x > u\}\big(F_0(u) - F_0(t_0)\big)}{(1 - F_0(u))^2} \right\}.$$

We have:

$$P g_n(\cdot,\cdot\,;t) \sim \tfrac{1}{2} \frac{f_0(t_0)g(t_0)}{F_0(t_0)\big(1 - F_0(t_0)\big)}\, t^2, \text{ as } n \to \infty, \tag{5.6}$$

uniformly for t in a bounded interval $[0, M]$. Using the representation (5.5), it is easily checked that the process $U_n^{(0)}$ satisfies the following type of stochastic equicontinuity condition: for each $\epsilon > 0$, $\eta > 0$, and $M > 0$, there exists a $\delta > 0$ such that

$$\limsup_{n \to \infty} I\!\!P \left\{ \sup_{0 \le t, t' \le M, |t-t'| \le \delta} \big| U_n^{(0)}(t) - U_n^{(0)}(t') \big| > \eta \right\} < \epsilon. \tag{5.7}$$

(a "tightness condition"). Moreover, the variance of $U_n^{(0)}(t)$ converges, for $t > 0$, to

$$\frac{g(t_0)}{F_0(t_0)\big(1 - F_0(t_0)\big)}\, t,$$

uniformly for t in bounded intervals. It now follows from (5.6) and (5.7) that the process $U_n^{(0)}$, restricted to $[0, \infty)$, converges in distribution to the process U, restricted to $[0, \infty)$ (in the topology of uniform convergence on compacta). Since a similar line of argument holds for $U_n^{(0)}$, restricted to $(-\infty, 0]$, the result follows. $\square$

Remark 5.2. Lemma 5.1 can also be proved by a martingale argument. Note, for example, that $\left\{ n^{2/3}\big(W_n^{(0)}(t_0 + n^{-1/3}t) - W_n^{(0)}(t_0)\big) : t \ge 0 \right\}$ is a martingale. For negative values of t, we can use a martingale with time running backward.

We now define, for each $a > 0$, the random variable $T_n^{(0)}(a)$ by

$$T_n^{(0)}(a) = \sup\{t \in [0, T_{(n)}] : V_n^{(0)}(t) - \big(a - F_0(t_0)\big)G_n^{(0)}(t) \text{ is minimal}\}. \tag{5.8}$$

Then the process

$$\left\{ \big(G_n^{(0)}(T_n^{(0)}(a)), V_n^{(0)}(T_n^{(0)}(a)) + F_0(t_0)G_n^{(0)}(T_n^{(0)}(a)) \big) : a \in (0,1) \right\}$$

runs through vertices of the cumulative sum diagram S_n, consisting of the points

$$P_j = \big(G_n^{(0)}(T_{(j)}), V_n^{(0)}(T_{(j)}) + F_0(t_0)G_n^{(0)}(T_{(j)}) \big), 1 \le j \le n,$$

and $P_0 = (0,0)$.

In a similar way, we define the process $\{T(a) : a \in \mathbb{R}\}$ by

$$T(a) = \sup\left\{t \in \mathbb{R} : U(t) - a \cdot \frac{g(t_0)t}{F_0(t_0)(1 - F_0(t_0))} \text{ is minimal}\right\}, \qquad (5.9)$$

where the process U is defined by (5.5). The processes $T_n^{(0)}$ and T are similar to processes studied in Groeneboom (1988) and (1989). They run through the locations of the vertices of the convex minorant of respectively, the process $V_n^{(0)}$ (in the "time scale" $G_n^{(0)}$) and the process U. We now show that a rescaled version of the process $T_n^{(0)}$ converges in distribution (in the space $D(\mathbb{R})$ with the Skorohod topology) to the process T.

Lemma 5.2. Let the conditions of Lemma 5.1 be satisfied, and let $a_0 = F_0(t_0)$. Moreover, let $T_n^{(0)}(a)$ be defined by (5.8) for $a \in (0,1)$, and let $T_n^{(0)}(a) = 0$, $a \leq 0$, and $T_n^{(0)}(a) = T_{(n)}, a \geq 1$. Then the process

$$\left\{n^{1/3}\{T_n^{(0)}(a_0 + n^{-1/3}a) - t_0\} : a \in \mathbb{R}\right\} \qquad (5.10)$$

converges distribution, in the space $D(\mathbb{R})$ with the Skorohod topology, to the process $\{T(a) : a \in \mathbb{R}\}$, defined by (5.9).

Proof. By a simple scaling argument, it is seen that

$$n^{1/3}\{T_n^{(0)}(a_0 + n^{-1/3}a) - t_0\}$$
$$= \sup\left\{t : U_n^{(0)}(t) - a \cdot n^{1/3}\{G_n^{(0)}(t_0 + n^{-1/3}t) - G_n^{(0)}(t_0)\} \text{ is minimal}\right\}.$$

Since, with probability one,

$$\lim_{n \to \infty} n^{1/3}\left\{G_n^{(0)}(t_0 + n^{-1/3}t) - G_n^{(0)}(t_0)\right\} = \frac{g(t_0)t}{F_0(t_0)(1 - F_0(t_0))},$$

uniformly for t in bounded intervals, the process

$$\left\{U_n^{(0)}(t) - a \cdot n^{1/3}\{G_n^{(0)}(t_0 + n^{-1/3}t) - G_n^{(0)}(t_0)\} : (t,a) \in \mathbb{R}^2\right\} \qquad (5.11)$$

converges in distribution (in the topology of uniform convergence on compacta), to the process

$$\left\{U(t) - a \cdot \frac{g(t_0)t}{F_0(t_0)(1 - F_0(t_0))} : (t,a) \in \mathbb{R}^2\right\}. \qquad (5.12)$$

It is proved below (Lemma 5.3) that for each $\epsilon > 0$ and $M_1 > 0$ an $M_2 > 0$ can be found such that

$$\mathbb{P}\left\{\sup_{a \in [-M_1, M_1]} n^{1/3}\big|T_n^{(0)}(a_0 + n^{-1/3}a) - t_0\big| > M_2\right\} < \epsilon, \qquad (5.13)$$

for all sufficiently large n. The process $\{T(a) : a \in I\!R\}$ is a function of the process (5.12), and it is shown in Groeneboom (1989) that $\{T(a) : a \in I\!R\}$ is an increasing Markovian jump process, with almost surely a finite number of jumps in each finite interval and that $\{T(a) - f_0(t_0)^{-1}a : a \in I\!R\}$ is a stationary process. Therefore, by (5.13) and an almost sure construction, similar to Theorem 2.7 in Kim and Pollard (1990) (which is a form of the continuous mapping theorem that can be used for "argmax functionals"), it follows that the process (5.10), considered as a function of the process (5.11), converges in distribution (in the Skorohod topology) to the corresponding function of the process (5.12). However, the latter function is just the process $\{T(a) : a \in I\!R\}$.   □

The following lemma establishes (5.13).

Lemma 5.3. For each $\epsilon > 0$ and $M_1 > 0$ an $M_2 > 0$ can be found such that

$$I\!P\left\{ \max_{a \in [-M_1, M_1]} n^{1/3}\{T_n^{(0)}(a_0 + n^{-1/3}a) - t_0\} > M_2 \right\} < \epsilon,$$

and

$$I\!P\left\{ \min_{a \in [-M_1, M_1]} n^{1/3}\{T_n^{(0)}(a_0 + n^{-1/3}a) - t_0\} < -M_2 \right\} < \epsilon,$$

for all sufficiently large n.

Proof. We only prove the first inequality, since the second inequality can be proved in a completely analogous way. First note that

$$I\!P\left\{ \max_{a \in [-M_1, M_1]} n^{1/3}\{T_n^{(0)}(a_0 + n^{-1/3}a) - t_0\} > M_2 \right\}$$
$$= I\!P\left\{ n^{1/3}\{T_n^{(0)}(a_0 + n^{-1/3}M_1) - t_0\} > M_2 \right\}$$

since the process $T_n^{(0)}$ is nondecreasing. Furthermore,

$$I\!P\left\{ n^{1/3}\{T_n^{(0)}(a_0 + n^{-1/3}M_1) - t_0\} > M_2 \right\}$$
$$\leq I\!P\left\{ U_n^{(0)}(t) - n^{1/3}M_1\{G_n^{(0)}(t_0 + n^{-1/3}t) - G_n^{(0)}(t_0)\} \right.$$
$$\leq 0, \text{ for some } t > M_2 \right\}.$$

We have

$$n^{-2/3}U_n^{(0)}(t) - n^{-1/3}M_1 \cdot \{G_n^{(0)}(t_0 + n^{-1/3}t) - G_n^{(0)}(t_0)\}$$
$$= V_n^{(0)}(t_0 + n^{-1/3}t) - V_n^{(0)}(t_0) - n^{-1/3}M_1\{G_n^{(0)}(t_0 + n^{-1/3}t) - G_n^{(0)}(t_0)\}$$
$$= W_n^{(0)}(t_0 + n^{-1/3}t) - W_n^{(0)}(t_0)$$
$$+ \int_{[t_0, t_0 + n^{-1/3}t]} \{F_0(t') - F_0(t_0) - n^{-1/3}M_1\} dG_n^{(0)}(t').$$

The process $\{W_n^{(0)}(t_0 + u) - W_n^{(0)}(t_0) : u \geq 0\}$, is a martingale with respect to the self-induced filtration $\{\mathcal{F}_u : u \geq 0\}$, where

$$\mathcal{F}_u = \sigma\{W_n^{(0)}(t_0 + u') - W_n^{(0)}(t_0) : 0 \leq u' \leq u\}.$$

Let $Z_n^{(0)}(u) = \left\{ W_n^{(0)}(t_0+u) - W_n^{(0)}(t_0) \right\}^2$, then $\{ Z_n^{(0)}(u) : u \geq 0 \}$ is a submartingale and, by Doob's inequality:

$$\mathbb{P}\left\{ \sup_{0 \leq u \leq u_0} Z_n^{(0)}(u) \right\} \leq 4\mathbb{P}Z_n^{(0)}(u_0) = 4n^{-1} \left\{ \frac{g(t_0)}{F_0(t_0)\left(1 - F_0(t_0)\right)} u_0 + \mathcal{O}(u_0^2) \right\},$$

if u_0 satisfies $F_0(t_0 + u_0) < 1$. So in particular we get, for $\epsilon > 0$ and $A > 0$:

$$\mathbb{P}\Big\{ \exists u \in [(j-1)n^{-1/3}, jn^{-1/3}) : n^{2/3}\big| W_n^{(0)}(t_0 + u) - W_n^{(0)}(t_0) \big|$$
$$> \epsilon(j-1)^2 + A \Big\}$$
$$\leq 4n^{4/3}\mathbb{P}Z_n^{(0)}(jn^{-1/3}) / \left\{ \epsilon(j-1)^2 + A \right\}^2$$
$$\leq 4n^{4/3}\left\{ n^{-1}c \cdot jn^{-1/3} \right\} / \left\{ \epsilon(j-1)^2 + A \right\}^2 = c \cdot j / \left\{ \epsilon(j-1)^2 + A \right\}^2,$$

for some constant $c > 0$, if $0 \leq jn^{-1/3} \leq u_0$, where c does not depend on $jn^{-1/3}$, for $jn^{-1/3} \leq u_0$.

By the same arguments as used in Lemma 4.1 in Kim and Pollard (1990), we get from this that for each $\epsilon > 0$ there exist random variables A_n of order $\mathcal{O}_p(1)$ such that

$$\big| W_n^{(0)}(t_0 + u) - W_n^{(0)}(t_0) \big| \leq \epsilon u^2 + n^{-2/3}A_n, \text{ if } 0 \leq u \leq u_0. \tag{5.14}$$

By the fact that F_0 is differentiable in t_0, with a strictly positive derivative $f_0(t_0)$, there exist, for each $M > 0$ and $\eta > 0$ an $M_2 > 0$ and $\epsilon > 0$ such that

$$\int_{[t_0, t_0+u]} \left\{ F_0(t') - F_0(t_0) - n^{-1/3}M_1 \right\} dG_n^{(0)}(t')$$
$$\geq \max\{ Mn^{-2/3}, \epsilon u^2 \}, \forall u \in [M_2 n^{-1/3}, u_0], \tag{5.15}$$

with probability bigger than $1 - \eta$. Combining (5.14) and (5.15) we get

$$\mathbb{P}\Big\{ U_n^{(0)}(t) - n^{1/3}M_1\big\{ G_n^{(0)}(t_0 + n^{-1/3}t) - G_n^{(0)}(t_0) \big\} < 0, \text{ for some } t, M_2$$
$$\leq t \leq u_0 n^{1/3} \Big\} \leq \mathbb{P}\{ A_n > M \} + \eta$$

and the last expression can be made smaller than 2η, by taking M sufficiently large.

Finally, we have to deal with the behavior of the process for values of $t \geq u_0 n^{1/3}$, where u_0 is chosen in such a way that $F_0(t_0 + u_0)$ is close to 1. But for these values of t the process

$$U_n^{(0)}(t) - n^{1/3}M_1\big\{ G_n^{(0)}(t_0 + n^{-1/3}t) - G_n^{(0)}(t_0) \big\}, \, t \geq 0,$$

will be increasing for all sufficiently large n, since negative terms of the form

$$-1/\left\{ n\big(1 - F_0(T_{(i)})\big) \right\}$$

in the martingale part of the process will be compensated by terms of the form

$$n^{-1}\{F_0(T_{(i)}) - F_0(t_0) - n^{-1/3}M_1\} / \{F_0(T_{(i)})(1 - F_0(T_{(i)}))\}^2,$$

in the "drift" part of the process.   □

The distribution of the 1-step estimator $F_n^{(1)}$ at t_0 can now be found by using the following relation between $F_n^{(1)}$ and $T_n^{(0)}$:

$$\mathbb{P}\{F_n^{(1)}(t_0) - F_0(t_0) > x\} = \mathbb{P}\{T_n^{(0)}(a_0 + x) < t_0\}. \tag{5.16}$$

Relation (5.16) is easily verified by drawing a picture of the situation. The preceding results yield the limiting distribution of $F_n^{(1)}(t_0)$.

Theorem 5.2. For each $x \in \mathbb{R}$ we have

$$\lim_{n \to \infty} \mathbb{P}\left\{n^{1/3}\{F_n^{(1)}(t_0) - F_0(t_0)\} > x\right\} = \mathbb{P}\{T(0) > c \cdot x\},$$

where $T(0)$ is defined by (5.9), and where

$$c = f_0(t_0)^{-1}. \tag{5.17}$$

Proof. Using (5.16), we only have to consider

$$\mathbb{P}\left\{T_n^{(0)}(a_0 + xn^{-1/3}) < t_0\right\}.$$

But we have, by Lemma 5.2, for fixed $x \in \mathbb{R}$,

$$\mathbb{P}\left\{T_n^{(0)}(a_0 + xn^{-1/3}) < t_0\right\}$$
$$= \mathbb{P}\left\{n^{1/3}\{T_n^{(0)}(a_0 + xn^{-1/3}) - t_0\} < 0\right\} \to \mathbb{P}\{T(x) < 0\},$$

as $n \to \infty$. Using the stationarity of the process $\{T(a) - c \cdot a : a \in \mathbb{R}\}$, proved in Groeneboom (1989), we get

$$\mathbb{P}\{T(x) < 0\} = \mathbb{P}\{T(x) - c \cdot x < -c \cdot x\} = \mathbb{P}\{T(0) < -c \cdot x\}.$$

The statement now follows from the fact that $T(0)$ has a density which is symmetric about zero.   □

We will now derive the asymptotic distribution of $\hat{F}_n(t_0)$, where $\hat{F}_n$ is the NPMLE. We first show that the distance between $\hat{F}_n(t)$ and $F_0(t_0)$ is $\mathcal{O}_p(n^{-1/3})$, for t in an interval of the form $[t_0 - Mn^{-1/3}, t_0 + Mn^{-1/3}]$.

Lemma 5.4. For each $M > 0$ we have:

$$\sup_{t \in [-M, M]} |\hat{F}_n(t_0 + n^{-1/3}t) - F_0(t_0)| = \mathcal{O}_p(n^{-1/3}).$$

Proof. We first show that the probability that (for large n) $\hat{F}_n$ does not have a jump in an interval of the form $[t_0 - Mn^{-1/3}, t_0 + Mn^{-1/3}]$ can be made arbitrarily small by taking M sufficiently large. Specifically, we will show that for each $\epsilon > 0$, there exists an $M > 0$ such that

$$I\!\!P\left\{\hat{F}_n(t_0 - Mn^{-1/3}) \geq F_0(t_0)\right\} < \epsilon, \tag{5.18}$$

for all large n. Since it can be shown in a similar way that for each $\epsilon > 0$, there exists an $M > 0$ such that

$$I\!\!P\left\{\hat{F}_n(t_0 + Mn^{-1/3}) \leq F_0(t_0)\right\} < \epsilon, \tag{5.19}$$

for all large n, we get that

$$\hat{F}_n(t_0 - Mn^{-1/3}) < F_0(t_0) < \hat{F}_n(t_0 + Mn^{-1/3}),$$

with probability $\geq 1 - 2\epsilon$ for all large n and a suitably chosen M, implying that $\hat{F}_n$ has a jump in $[t_0 - Mn^{-1/3}, t_0 + Mn^{-1/3}]$.

For the proof of (5.18), we will assume, as in Proposition 1.1, that $\delta_{(1)} = 1$ and $\delta_{(n)} = 0$, since, by the assumptions on F_0 and G, we can always assume that (for large n) there exist observation times T_i and T_j such that $T_i < T_j$, $\delta_i = 1_{\{X_i \leq T_i\}} = 1$ and $\delta_j = 1_{\{X_j > T_j\}} = 0$ (see the discussion in the first paragraph of the proof of Proposition 1.2).

Let τ_n be the last jump time of $\hat{F}_n$ before $t_0 - Mn^{-1/3}$, i.e.,

$$\tau_n = \max\{t \leq t_0 - Mn^{-1/3} : \hat{F}_n(t-) \neq \hat{F}_n(t)\}.$$

By the assumption just made, τ_n is, for each large n, the maximum over a non-empty set, since $T_{(1)}$ belongs to that set for all large n. By Proposition 1.1, we must have

$$\sum_{\tau_n \leq T_i < T_{(j)}} \left\{\frac{\delta_i}{\hat{F}_n(T_i)} - \frac{1 - \delta_i}{1 - \hat{F}_n(T_i)}\right\} \geq 0,$$

for all $T_{(j)} > \tau_n$. Hence, if $\hat{F}_n(t_0 - Mn^{-1/3}) \geq F_0(t_0)$, we get

$$\sum_{\tau_n \leq T_i < T_{(j)}} \left\{\frac{\delta_i}{F_0(t_0)} - \frac{1 - \delta_i}{1 - F_0(t_0)}\right\} \geq \sum_{\tau_n \leq T_i < T_{(j)}} \left\{\frac{\delta_i}{\hat{F}_n(T_i)} - \frac{1 - \delta_i}{1 - \hat{F}_n(T_i)}\right\} \geq 0, \tag{5.20}$$

for all $T_{(j)} > \tau_n$. But we have

$$\sum_{\tau_n \leq T_i < t_0 - \frac{1}{2}Mn^{-1/3}} \left\{\frac{\delta_i}{F_0(t_0)} - \frac{1 - \delta_i}{1 - F_0(t_0)}\right\}$$

$$= \sum_{\tau_n \leq T_i < t_0 - \frac{1}{2}Mn^{-1/3}} \left\{\frac{F_0(T_i)}{F_0(t_0)} - \frac{1 - F_0(T_i)}{1 - F_0(t_0)} + \frac{\delta_i - F_0(T_i)}{F_0(t_0)(1 - F_0(t_0))}\right\}. \tag{5.21}$$

The right-hand side of (5.21) can be written

$$\{F_0(t_0)(1 - F_0(t_0))\}^{-1} \sum_{\tau_n \leq T_i < t_0 - \frac{1}{2}Mn^{-1/3}} \{\delta_i - F_0(T_i) - (F_0(t_0) - F_0(T_i))\},$$

and by a similar technique as used to derive (5.14) (using a martingale with time running backward), we get

$$\mathbb{P}\left\{\sup_{t \leq t_0 - Mn^{-1/3}} \sum_{t \leq T_i \leq t_0 - \frac{1}{2}Mn^{-1/3}} \{\delta_i - F_0(T_i) - (F_0(t_0) - F_0(T_i))\} \geq 0\right\} < \epsilon,$$

for M and n large. This, however, contradicts (5.20), and (5.18) follows.

The remaining part of the proposition is proved in a similar way. Using the same arguments as above, one can show that, for large $M > 0$, $\hat{F}_n$ will have with high probability a jump in the interval

$$[t_0 - 2Mn^{-1/3}, t_0 - Mn^{-1/3}].$$

Let τ_n be such a jump time, and suppose that

$$\hat{F}_n(\tau_n) \leq F_0(t_0 - c \cdot n^{-1/3}),$$

where $c > 2M$. By Proposition 1.1 we have

$$\sum_{T_{(j)} < T_i \leq \tau_n} \left\{\frac{\delta_i}{\hat{F}_n(T_i)} - \frac{1 - \delta_i}{1 - \hat{F}_n(T_i)}\right\} \leq 0, \tag{5.22}$$

for all $T_{(j)} < \tau_n$. On the other hand, if $\hat{F}_n(\tau_n) \leq F_0(t_0 - c \cdot n^{-1/3})$, we get

$$\sum_{t_0 - c \cdot n^{-1/3} < T_i \leq \tau_n} \left\{\frac{\delta_i}{\hat{F}_n(T_i)} - \frac{1 - \delta_i}{1 - \hat{F}_n(T_i)}\right\}$$
$$\geq \sum_{t_0 - c \cdot n^{-1/3} < T_i \leq \tau_n} \left\{\frac{\delta_i}{F_0(t_0 - c \cdot n^{-1/3})} - \frac{1 - \delta_i}{1 - F_0(t_0 - c \cdot n^{-1/3})}\right\},$$

and

$$n^{-1/3} \sum_{t_0 - c \cdot n^{-1/3} < T_i \leq \tau_n} \left\{\frac{\delta_i}{F_0(t_0 - c \cdot n^{-1/3})} - \frac{1 - \delta_i}{1 - F_0(t_0 - c \cdot n^{-1/3})}\right\}$$
$$= n^{-1/3} \sum_{t_0 - c \cdot n^{-1/3} < T_i \leq \tau_n} \left\{\frac{F_0(T_i)}{F_0(t_0 - c \cdot n^{-1/3})} - \frac{1 - F_0(T_i)}{1 - F_0(t_0 - c \cdot n^{-1/3})}\right\}$$
$$+ n^{-1/3} \sum_{t_0 - c \cdot n^{-1/3} < T_i \leq \tau_n} \frac{\delta_i - F_0(T_i)}{F_0(t_0 - c \cdot n^{-1/3})(1 - F_0(t_0 - c \cdot n^{-1/3}))}. \tag{5.23}$$

The last term at the right-hand side of (5.23) has expectation zero and a variance which is asymptotically bounded above by

$$(c - M)g(t_0)/\{F_0(t_0)(1 - F_0(t_0))\}.$$

The first term at the right-hand side of (5.23) is (almost surely) asymptotically bounded below by

$$\tfrac{1}{2}(c - 2M)^2 f_0(t_0)g(t_0)/\{F_0(t_0)(1 - F_0(t_0)\}.$$

This implies that, with high probability, (5.23) will be strictly positive, for large n and large c (with c only depending on M). This contradicts (5.23). Hence, with high probability

$$\hat{F}_n(t_0 - Mn^{-1/3}) \geq \hat{F}_n(\tau_n) \geq F_0(t_0 - cn^{-1/3}). \tag{5.24}$$

Combining (5.18) and (5.24) we find that for each $\epsilon > 0$ there exist positive constants c and M such that

$$\mathbb{P}\{F_0(t_0 - cn^{-1/3}) \leq \hat{F}_n(t_0 - Mn^{-1/3}) \leq F_0(t_0)\} > 1 - \epsilon,$$

for all sufficiently large n. Likewise we get that for each $\epsilon > 0$ there exist positive constants c and M such that

$$\mathbb{P}\{F_0(t_0 + cn^{-1/3}) \geq \hat{F}_n(t_0 + Mn^{-1/3}) \geq F_0(t_0)\} > 1 - \epsilon.$$

The lemma now follows from the monotonicity of $\hat{F}_n$.   □

We can now give a proof of Theorem 5.1.

Proof of Theorem 5.1. Let $W_{\hat{F}_n}$, $G_{\hat{F}_n}$ and $V_{\hat{F}_n}$ be defined by (5.1) to (5.3), respectively, with F_0 everywhere replaced by $\hat{F}_n$. Then $\hat{F}_n$ can be characterized as the slope of the convex minorant of the self-induced cumulative sum diagram, consisting of the points

$$P_j = \left(G_{\hat{F}_n}(T_{(j)}), V_{\hat{F}_n}(T_{(j)})\right),$$

where $P_0 = (0,0)$ (see Proposition 1.4 in Chapter 1 of Part II, for a similar characterization of the NPMLE in the situation of interval censoring, Case 2).

Fix an interval $[t_0 - Mn^{-1/3}, t_0 + Mn^{-1/3}]$, where M is a positive constant, and let τ_n^- and τ_n^+ be the two points, corresponding to, respectively, the last change of slope $\leq t_0 - Mn^{-1/3}$ and the first change of slope $\geq t_0 + Mn^{-1/3}$ of the convex minorant of the cumulative sum diagram (this means that

$$\left(G_{\hat{F}_n}(\tau_n^-), V_{\hat{F}_n}(\tau_n^-)\right) \quad \text{and} \quad \left(G_{\hat{F}_n}(\tau_n^+), V_{\hat{F}_n}(\tau_n^+)\right)$$

are vertices of the convex minorant).

By Lemma 5.4 and the strict monotonicity of F_0 at t_0, we have $\tau_n^- - t_0 = \mathcal{O}_p(n^{-1/3})$ and likewise $\tau_n^+ - t_0 = \mathcal{O}_p(n^{-1/3})$.

The convex minorant of the process $V_{\hat{F}_n}$ coincides on the interval $(\tau_n^-, \tau_n^+]$ with the convex minorant of the process $V_{\hat{F}_n}$, *restricted to the interval* $(\tau_n^-, \tau_n^+]$. Moreover, also by Lemma 5.4, the process

$$
\begin{aligned}
t \mapsto \; & n^{2/3}\big\{ W_{\hat{F}_n}(t_0 + n^{-1/3}t) - W_{\hat{F}_n}(t_0) \big\} \\
& + n^{2/3} \int_{[0, t_0 + n^{-1/3}t]} \big(\hat{F}_n(t') - F_0(t_0) \big) \, dG_{\hat{F}_n}(t') \\
& - n^{2/3} \int_{[0, t_0]} \big(\hat{F}_n(t') - F_0(t_0) \big) \, dG_{\hat{F}_n}(t')
\end{aligned}
\tag{5.25}
$$

converges in distribution (in the topology of uniform convergence on compacta) to the process $\{U(t) : t \in \mathbb{R}\}$, defined by (5.4). This means that the process

$$
t \mapsto n^{1/3}\big\{ \hat{F}_n(t_0 + n^{-1/3}t) - F_0(t_0) \big\}
\tag{5.26}
$$

will converge in distribution (in the Skorohod topology) to the left derivative of the convex minorant of the process U, since, for each $M > 0$, jump times τ_n^- and τ_n^+ can be found, defined as above, such that the process (5.26) coincides on the interval $(-n^{1/3}(t_0 - \tau_n^-), n^{1/3}(\tau_n^+ - t_0)]$ with the left derivative of the convex minorant of the process (5.25), restricted to this interval. This shows that the process (5.26) has the same limiting distribution as the process

$$
t \mapsto n^{1/3}\big\{ F_n^{(1)}(t_0 + n^{-1/3}t) - F_0(t_0) \big\}.
\tag{5.27}
$$

Hence we get from Theorem 5.2:

$$
\lim_{n \to \infty} \mathbb{P}\Big\{ n^{1/3}\big\{ \hat{F}_n(t_0) - F_0(t_0) \big\} > x \Big\} = \mathbb{P}\{ T(0) > c \cdot x \},
$$

where $T(0)$ is defined by (5.9), and $c = f_0(t_0)^{-1}$ (see (5.17)). Thus

$$
\begin{aligned}
& \mathbb{P}\Big\{ n^{1/3}\big\{ \hat{F}_n(t_0) - F_0(t_0) \big\} \big/ \big\{ \tfrac{1}{2} F_0(t_0)(1 - F_0(t_0)) f_0(t_0)/g(t_0) \big\}^{1/3} > x \Big\} \\
& \to \mathbb{P}\Big\{ T(0) > x \big\{ \tfrac{1}{2} F_0(t_0)(1 - F_0(t_0)) \big/ \{ g(t_0) f_0(t_0)^2 \} \big\}^{1/3} \Big\}, \; n \to \infty.
\end{aligned}
\tag{5.28}
$$

Now $T(0)$ is the last time where the process

$$
t \mapsto W\Big(t \cdot g(t_0) \big/ \big(F_0(t_0)(1 - F_0(t_0)) \big) \Big) + \frac{1}{2} \frac{f_0(t_0) g(t_0)}{F_0(t_0)\big(1 - F_0(t_0)\big)} t^2
$$

reaches its minimum. By Brownian scaling (i.e., the property that $t \mapsto W(t)$ has the same distribution as $t \mapsto c^{-1/2} W(ct)$, for each constant $c > 0$), this means that

$$
\tfrac{1}{2} T(0) \Big/ \Big\{ \tfrac{1}{2} F_0(t_0)(1 - F_0(t_0)) \big/ \{ g(t_0) f_0(t_0)^2 \} \Big\}^{1/3}
\tag{5.29}
$$

is the last time where

$$t \mapsto W(t) + t^2$$

reaches its minimum. Using the symmetry of the distribution of Brownian motion with repect to the time axis, this means that (5.29) has the same distribution as the last time where

$$t \mapsto W(t) - t^2$$

reaches its maximum. The result now follows from (5.28).   □

5.2 Interval censoring, Case 2

We shall prove the following result for the 1-step estimator $F_n^{(1)}$.

Theorem 5.3. Let F_0 and H be differentiable at t_0 and (t_0, t_0), respectively, with strictly positive derivatives $f_0(t_0)$ and $h(t_0, t_0)$. By continuous differentiability of H at (t_0, t_0) is meant that the density $h(t, u)$ is continuous in (t, u) if $t < u$ and (t, u) is sufficiently close to (t_0, t_0) and that $h(t, t)$, defined by

$$h(t, t) = \lim_{u \downarrow t} h(t, u),$$

is continuous in t, for t in a neighborhood of t_0.

Let $0 < F_0(t_0), H(t_0, t_0) < 1$, and let $F_n^{(1)}$ be the estimator of F_0, obtained at the first step of the iterative convex minorant algorithm. Then

$$(n \log n)^{1/3} \{ F_n^{(1)}(t_0) - F_0(t_0) \} / \{ \tfrac{3}{4} f_0(t_0)^2 / h(t_0, t_0) \}^{1/3} \xrightarrow{\mathcal{D}} 2Z,$$

where Z is the last time where standard two-sided Brownian motion minus the parabola $y(t) = t^2$ reaches its maximum.

According to our working hypothesis, formulated at beginning of this chapter, this leads us to believe that the NPMLE has the same limiting behavior at t_0 as $F_n^{(1)}$. Proceeding as before, we introduce the processes $W_n^{(0)}$ and $G_n^{(0)}$ defined by

$$W_n^{(0)} = W_{F_0}, \text{ and } G_n^{(0)} = G_{F_0},$$

where W_{F_0} and G_{F_0} are defined by (1.25) and (1.29), respectively. The process $V_n^{(0)}$ is defined as in (5.3). We have the following result for $V_n^{(0)}$.

Lemma 5.5. Let $\delta_n = (n \log n)^{-1/3}$ and let the process $U_n^{(0)}$ be defined by

$$U_n^{(0)}(t) = \delta_n^{-2} (\log n)^{-1} \{ V_n^{(0)}(t_0 + \delta_n t) - V_n^{(0)}(t_0) \}, \ t \in \mathbb{R},$$

where $U_n^{(0)}(t) = 0$, if $t \leq -t_0 \delta_n^{-1}$. Then $U_n^{(0)}$ converges in distribution, in the topology of uniform convergence on compacta on the space of locally bounded real-valued functions on $\mathbb{R}$, to the process U, defined by

$$U(t) = \sqrt{\tfrac{2}{3}h(t_0, t_0)/f_0(t_0)}\, W(t) + \tfrac{1}{3}h(t_0, t_0)t^2,\ t \in \mathbb{R}, \qquad (5.30)$$

where W is (standard) two-sided Brownian motion on $\mathbb{R}$, originating from zero.

Proof. We first show that the process

$$t \mapsto \delta_n^{-2}(\log n)^{-1}\big\{W_n^{(0)}(t_0 + \delta_n t) - W_n^{(0)}(t_0)\big\},\ t \geq 0 \qquad (5.31)$$

converges, in the topology of uniform convergence on compacta, to the process

$$t \mapsto \sqrt{\tfrac{2}{3}h(t_0, t_0)/f_0(t_0)}\, W(t),\ t \geq 0. \qquad (5.32)$$

It is clear that the process (5.31) is not a martingale, but by removing some terms, it can be made a martingale. Fix $M > 0$ and define the sets of observation times A_n and B_n by

$$\begin{aligned}
A_n = \ & \{T_i \in (t_0, t_0 + M\delta_n] : U_i - T_i > M\delta_n\} \\
& \cup \{U_i \in (t_0, t_0 + M\delta_n] : U_i - T_i > M\delta_n\},
\end{aligned} \qquad (5.33)$$

and

$$B_n = \{T_i \in (t_0, t_0 + M\delta_n] : T_i \notin A_n\} \cup \{U_i \in (t_0, t_0 + M\delta_n] : U_i \notin A_n\}. \qquad (5.34)$$

Furthermore, define the processes $W_{n,1}$ and $W_{n,2}$ on $[0, M]$ by

$$\begin{aligned}
W_{n,1}(t) &= \delta_n^{-2}(\log n)^{-1}\int_{t' \in (t_0, t_0 + \delta_n t],\, t' \in A_n} dW_n^{(0)}(t'), \\
W_{n,2}(t) &= \delta_n^{-2}(\log n)^{-1}\int_{t' \in (t_0, t_0 + \delta_n t],\, t' \in B_n} dW_n^{(0)}(t'),
\end{aligned}$$

for $t \in [0, M]$. Then we get

$$\max_{t \in [0, M]} |W_{n,2}(t)| \leq \delta_n^{-2}(n \log n)^{-1} \sum_{T_{(i)} \in B_n} |\Delta W_{n,2}(T_{(i)})|,$$

where the $\Delta W_{n,2}(T_{(i)})$ are the jumps of the process $W_{n,2}$. Computing expectations and using Markov's inequality, we obtain, for $\epsilon > 0$,

$$\begin{aligned}
& \mathbb{P}\big\{\max_{t \in [0,M]} |W_{n,2}(t)| > \epsilon\big\} \\
& \leq 2\epsilon^{-1}\delta_n^{-2}(\log n)^{-1}\int_{t, u \in (t_0, t_0 + M\delta_n]} dH(t, u) = \mathcal{O}\big((\log n)^{-1}\big).
\end{aligned}$$

On the other hand, the process $t \mapsto W_{n,1}(t)$ is a martingale, and a straightforward computation shows that the variance of $W_{n,1}(t)$ is given by

$$\begin{aligned}
& n\delta_n^2 \int_{t' \in (t_0, t_0 + t\delta_n],\, u - t' > M\delta_n} \left\{\frac{1}{F_0(t')} + \frac{1}{F_0(u) - F_0(t')}\right\} dH(t', u) \\
& + n\delta_n^2 \int_{u \in (t_0, t_0 + t\delta_n],\, u - t' > M\delta_n} \left\{\frac{1}{F_0(u) - F_0(t')} + \frac{1}{1 - F_0(u)}\right\} dH(t', u),
\end{aligned}$$

which is asymptotically equivalent to

$$\tfrac{2}{3}h(t_0,t_0)/f_0(t_0)\,t,\ t>0.$$

This shows that the process (5.31) converges to the Wiener process (5.32). Since a similar argument holds for negative values of t, we get that the process

$$t \mapsto \delta_n^{-2}(\log n)^{-1}\big\{W_n^{(0)}(t_0+\delta_n t) - W_n^{(0)}(t_0)\big\},$$

converges, in the topology of uniform convergence on compacta, to the process

$$t \mapsto \sqrt{\tfrac{2}{3}h(t_0,t_0)/f_0(t_0)}\,W(t),\ t \in \mathbb{R}.$$

Since, for each $M>0$

$$\mathbb{P}\big\{T_i < X_i \le U_i,\ T_i, U_i \in [t_0, t_0+M\delta_n]\big\} \sim \tfrac{1}{6}f_0(t_0)h(t_0,t_0)M^3\delta_n^3,$$

we get

$$\mathbb{P}\big\{T_i < X_i \le U_i,\ T_i, U_i \in [t_0, t_0+M\delta_n],\ \text{for some } i\big\}$$
$$\le n\mathbb{P}\big\{T_1 < X_1 \le U_1,\ T_1, U_1 \in [t_0, t_0+M\delta_n]\big\}$$
$$\sim \tfrac{1}{6}h(t_0,t_0)M^3(\log n)^{-1} \to 0,\ \text{as } n \to \infty.$$

Furthermore, with probability one,

$$\delta_n^{-2}(\log n)^{-1}\int_{t'\in[t_0,t_0+\delta_n t],\,t'<x\le u, u-t'>M\delta_n} \frac{F_0(t')-F_0(t_0)}{\big(F_0(u)-F_0(t')\big)^2}\,dP_n(x,t',u)$$
$$\to \tfrac{1}{6}h(t_0,t_0)t^2,$$

and likewise

$$\delta_n^{-2}(\log n)^{-1}\int_{u\in[t_0,t_0+\delta_n t],\,t'<x\le u, u-t'>M\delta_n} \frac{F_0(u)-F_0(t_0)}{\big(F_0(u)-F_0(t')\big)^2}\,dP_n(x,t',u)$$
$$\to \tfrac{1}{6}h(t_0,t_0)t^2,$$

uniformly for t in a bounded interval $[0, M]$, as $n \to \infty$. Since similar relations hold for $t<0$, the result now follows.   □

Next we define, in analogy with (5.8), for each $a>0$, the random variable $T_n^{(0)}(a)$ by

$$T_n^{(0)}(a) = \sup\{t \in J_n :\ V_n^{(0)}(t) - (a - F_0(t_0))G_n^{(0)}(t) \text{ is minimal}\},$$

where the index set J_n is defined as in Definition 1.1. Likewise, in analogy with (5.9), we define the process $\{T(a) : a \in \mathbb{R}\}$ by

$$T(a) = \sup\left\{t \in \mathbb{R} :\ U(t) - a \cdot \frac{2h(t_0,t_0)t}{3f_0(t_0)} \text{ is minimal}\right\}. \tag{5.35}$$

Lemma 5.2 now leads us to expect that

$$\left\{\delta_n^{-1}\big\{T_n^{(0)}(a_0+\delta_n a) - t_0\big\} : a \in \mathbb{R}\right\}$$

will converge in distribution to the process $\{T(a) : a \in \mathbb{R}\}$. This is indeed the case. The proof is similar to the proof of Lemma 5.2, where the (crucial) Lemma 5.3 is replaced by the following lemma.

Lemma 5.6. For each $\epsilon > 0$ and $M_1 > 0$ an $M_2 > 0$ can be found such that

$$\mathbb{P}\left\{\max_{a \in [-M_1, M_1]} \delta_n^{-1}\{T_n^{(0)}(a_0 + \delta_n a) - t_0\} > M_2\right\} < \epsilon,$$

and

$$\mathbb{P}\left\{\min_{a \in [-M_1, M_1]} \delta_n^{-1}\{T_n^{(0)}(a_0 + \delta_n a) - t_0\} < -M_2\right\} < \epsilon,$$

for all sufficiently large n.

Proof. We (again) only prove the first inequality. First note that

$$\mathbb{P}\left\{\max_{a \in [-M_1, M_1]} \delta_n^{-1}\{T_n^{(0)}(a_0 + \delta_n a) - t_0\} > M_2\right\}$$
$$= \mathbb{P}\left\{\delta_n^{-1}\{T_n^{(0)}(a_0 + \delta_n M_1) - t_0\} > M_2\right\}.$$

Furthermore,

$$\mathbb{P}\left\{\delta_n^{-1}\{T_n^{(0)}(a_0 + \delta_n M_1) - t_0\} > M_2\right\}$$
$$\leq \mathbb{P}\left\{U_n^{(0)}(t) - (\delta_n \log n)^{-1} M_1 \{G_n^{(0)}(t_0 + \delta_n t)\right.$$
$$\left. - G_n^{(0)}(t_0)\} \leq 0, \text{ for some } t > M_2\right\}.$$

Similarly as in the proof of Lemma 5.3, we have

$$\delta_n^2 (\log n) U_n^{(0)}(t) - \delta_n M_1 \cdot \{G_n^{(0)}(t_0 + \delta_n t) - G_n^{(0)}(t_0)\}$$
$$= V_n^{(0)}(t_0 + \delta_n t) - V_n^{(0)}(t_0) - \delta_n M_1 \{G_n^{(0)}(t_0 + \delta_n t) - G_n^{(0)}(t_0)\}$$
$$= W_n^{(0)}(t_0 + \delta_n t) - W_n^{(0)}(t_0)$$
$$+ \int_{[t_0, t_0 + \delta_n t]} \{F_0(t') - F_0(t_0) - \delta_n M_1\} dG_n^{(0)}(t').$$

Let the process X_n be defined by

$$X_n(t) = \delta_n^{-2}(\log n)^{-1}\left\{W_n^{(0)}(t_0 + t) - W_n^{(0)}(t_0)\right\}, \ t \geq 0.$$

We first derive an upper bound for the probabilities

$$\mathbb{P}\{|X_n(j\delta_n)| > \epsilon j^2 + m\}, j = 1, 2, \ldots,$$

where ϵ and m are arbitrarily chosen positive numbers.

Let the sets of observation times $A_{n,j}$ and $B_{n,j}$ be defined by

$$A_{n,j} = \left\{T_i \in (t_0, t_0 + j\delta_n] : F_0(U_i) - F_0(T_i) > \delta_n\right\}$$
$$\cup \{U_i \in (t_0, t_0 + j\delta_n] : F_0(U_i) - F_0(T_i) > \delta_n\},$$
$$B_{n,j} = \left\{T_i \in (t_0, t_0 + j\delta_n] : T_i \notin A_{n,j}\right\} \cup \left\{U_i \in (t_0, t_0 + j\delta_n] : U_i \notin A_{n,j}\right\}.$$

Furthermore, let the random variables $X'_{n,j}$ and $X''_{n,j}$ be defined by

$$X'_{n,j} = \int_{t\in(0,j\delta_n],\, t\in A_{n,j}} dX_n(t),$$

and

$$X''_{n,j} = \int_{t\in(0,j\delta_n],\, t\in B_{n,j}} dX_n(t).$$

We have:

$$
\begin{aligned}
I\!\!P\{|X'_{n,j}| > \tfrac{1}{2}(\epsilon j^2 + m)\} &\leq 4I\!\!P\big(X'_{n,j}(j\delta_n)\big)^2/(\epsilon j^2 + m)^2 \\
&\leq c \cdot n^{-1}\delta_n^{-4}(\log n)^{-1}\{j\delta_n\}/(\epsilon j^2 + m)^2 \\
&= c \cdot j/(\epsilon j^2 + m)^2,
\end{aligned}
$$

and

$$
\begin{aligned}
I\!\!P\{|X''_{n,j}| > \tfrac{1}{2}(\epsilon j^2 + m)\} &\leq 2I\!\!P|(X''_{n,j}|/(\epsilon j^2 + m) \\
&\leq c \cdot \delta_n^{-2}(\log n)^{-1}\delta_n^2/(\epsilon j^2 + m) \\
&= c/\{(\epsilon j^2 + m)\log n\},
\end{aligned}
$$

for some constant $c > 0$, not depending on j, if $j\delta_n$ satisfies $j\delta_n \leq \delta$, and $\delta > 0$ is chosen in such a way that $t \mapsto f_0(t)$ and $t \mapsto h(t,t)$ satisfy the local positivity and continuity conditions of Theorem 5.3 in the neighborhood $[t_0, t_0 + \delta]$. Moreover, we use the fact that, if $t_0 < T_i < U_i \leq t_0 + j\delta_n$, terms of the form $\pm\{T_i < X_i \leq U_i\}/\{F_0(U_i) - F_0(T_i)\}$ will give no contribution to $X'_{n,j}$ or $X''_{n,j}$, since these terms will occur with opposite signs and will cancel in the summation.

We also have:

$$
\begin{aligned}
I\!\!P\big\{\sup_{t\in[0,\delta_n]} |W_{n,1}(t)| > \tfrac{1}{2}(\epsilon j^2 + m)\big\} &\leq 4I\!\!P\big(W_{n,1}(\delta_n)\big)^2/(\epsilon j^2 + m)^2 \\
&\leq c/(\epsilon j^2 + m)^2,
\end{aligned}
$$

and

$$I\!\!P\Big\{\sup_{t\in[0,\delta_n]} |W_{n,2}(t)| > \tfrac{1}{2}(\epsilon j^2 + m)\Big\} \leq c/\{(\epsilon j^2 + m)\log n\},$$

for some constant $c > 0$, where $W_{n,1}$ and $W_{n,2}$ are defined as in the proof of Lemma 5.5, with $M = 1$. Hence

$$I\!\!P\Big\{\sup_{t\in[0,\delta_n]} |X_n(t)| > \tfrac{1}{2}(\epsilon j^2 + m)\Big\} \leq c/(\epsilon j^2 + m)^2 + c/\{(\epsilon j^2 + m)\log n\}.$$

It is clear that we get the same kind of upper bound for

$$I\!\!P\Big\{\sup_{t\in((j-1)\delta_n,j\delta_n]} \big|X_n(t) - X_n((j-1)\delta_n)\big| > \tfrac{1}{2}(\epsilon j^2 + m)\Big\},$$

if $j\delta_n \leq \delta$. Combining these results, we obtain that for $\epsilon > 0$ and $m > 0$:

$$
\begin{aligned}
I\!\!P\{\exists t \in ((j-1)\delta_n, j\delta_n] &: |X_n(t)| > \epsilon(j.-1)^2 + m\} \\
&\leq c \cdot j/\{\epsilon(j-1)^2 + m\}^2 + c/\{(\epsilon(j-1)^2 + m)\log n\},
\end{aligned}
$$

if $j\delta_n \leq \delta$. As in the proof of Lemma 5.3 (see (5.14)), this yields

$$\delta_n^2 |X_n(t)| \leq \epsilon t^2 + \delta_n^2 R_n, \text{ if } 0 \leq t \leq \delta,$$

where $R_n = \mathcal{O}_p(1)$.

As in (5.15), there exist, for each $M > 0$ and $\eta > 0$ an $M_2 > 0$ and $\epsilon > 0$ such that

$$(\log n)^{-1} \int_{[t_0, t_0+u]} \{F_0(t') - F_0(t_0) - \delta_n M_1\} dG_n^{(0)}(t') \geq \max\{M\delta_n^2, \epsilon u^2\},$$

$$\forall u \in [M_2\delta_n, \delta],$$

with probability bigger than $1 - \eta$. This is easily checked by taking a constant $k > 0$ such that

$$F_0(k\delta_n) - F_0(t_0) \geq \delta_n M_1,$$

and by splitting the points of increase of $G_n^{(0)}$ on the interval $[t_0, t_0 + k\delta_n]$ into two sets A_n and B_n, as in (5.33) and (5.34), with M replaced by constant k. On the interval $[t_0, t_0 + k\delta_n]$ we replace the integral by an integral over points T_i and U_i, belonging to the set A_n, and use that the probability of having an observation point in B_n, giving a non-zero contribution to the integral, will tend to zero, as $n \to \infty$. On the remaining interval we also remove points T_i and U_i such that $U_i - T_i < \delta_n$, showing that, with a probability tending to one, the integral

$$(\log n)^{-1} \int_{[t_0, t_0+u]} \{F_0(t') - F_0(t_0) - \delta_n M_1\} dG_n^{(0)}(t')$$

is bounded below by a term which is asymptotically equivalent to

$$\frac{2}{3} \int_{[t_0, t_0+u]} \{F_0(t') - F_0(t_0) - \delta_n M_1\} \frac{h(t,t)}{f_0(t)} dt.$$

Combining the preceding results, we obtain

$$\mathbb{P}\Big\{U_n^{(0)}(t) - \delta_n^{-1} M_1 (\log n)^{-1} \{G_n^{(0)}(t_0 + \delta_n t) - G_n^{(0)}(t_0)\} \leq 0,$$

$$\text{for some } t, M_2\delta_n < t \leq \delta\Big\}$$

$$\leq \mathbb{P}\{A_n > M\} + \eta,$$

and the last expression can be made smaller than 2η, by taking M sufficiently large.

Finally, we have to deal with the interval $[t_0 + \delta, \infty)$. But on this interval we will only get jumps downward, if

$$-\Delta W_n^{(0)}(T_{(i)}) \geq \{F_0(T_{(i)}) - F_0(t_0) - \delta_n M_1\} \Delta G_n^{(0)}(T_{(i)}), \qquad (5.36)$$

where $T_{(i)}$ is an observation point in $[t_0 + \delta, \infty)$, and where $\Delta W_n^{(0)}(T_{(i)})$ and $\Delta G_n^{(0)}(T_{(i)})$ are jumps of the processes $W_n^{(0)}$ and $G_n^{(0)}$, respectively. But this can only happen if

$$n\big|\Delta W_n^{(0)}(T_{(i)})\big| \leq \big\{F_0(t_0 + \delta) - F_0(t_0) - \delta_n M_1\big\}^{-1},$$

implying that the sum of the terms $\Delta W_n^{(0)}(T_{(i)})$, satisfying (5.36), for $T_{(i)} \in [t_0 + \delta, \infty)$, is bounded below by a fixed constant. But since, for some constant c, $0 < c < 1$,

$$\int_{[t_0, t_0 + \delta]} \big\{F_0(t) - F_0(t_0) - \delta_n M_1\big\} dG_n^{(0)}(t) > c\big|W_n^{(0)}(t_0 + \delta) - W_n^{(0)}(t_0)\big|$$

with a probability tending to one, and since

$$\int_{[t_0, t_0 + \delta]} \big\{F_0(t) - F_0(t_0) - \delta_n M_1\big\} dG_n^{(0)}(t)$$

will tend to ∞ with probability one, we will get

$$W_n^{(0)}(t_0 + t) - W_n^{(0)}(t_0) + \int_{[t_0, t_0 + t]} \big\{F_0(t') - F_0(t_0) - \delta_n M_1\big\} dG_n^{(0)}(t') > 0, \; \forall t \geq t_0 + \delta,$$

with probability tending to one.   □

Using Lemma 5.6, we can now give a proof of Theorem 5.3. The proof is quite similar to the proof of Theorem 5.2.

Proof of Theorem 5.3. First of all, similarly to (5.16), we have:

$$I\!\!P\{F_n^{(1)}(t_0) - F_0(t_0) > \delta_n a\} = I\!\!P\{T_n^{(0)}(a_0 + \delta_n a) < t_0\}.$$

Moreover, using Lemmas 5.5 and 5.6, we get that the process

$$\Big\{\delta_n^{-1}\big\{T_n^{(0)}(a_0 + \delta_n a) - t_0\big\} : a \in I\!\!R\Big\}$$

converges in the Skorohod topology on $D(I\!\!R)$ to the process $\{T(a) : a \in I\!\!R\}$, defined by (5.35). As before, the process

$$\{T(a) - f_0(t_0)^{-1}a : a \in I\!\!R\}$$

is stationary, since $T(a)$ is the last time where

$$t \mapsto \sqrt{\tfrac{2}{3}h(t_0, t_0)/f_0(t_0)}\, W(t) + \tfrac{1}{3}h(t_0, t_0)\big(t - f_0(t_0)^{-1}a\big)^2, \; t \in I\!\!R,$$

is minimal. Hence,

$$I\!\!P\Big\{\delta_n^{-1}\big\{F_n^{(1)}(t_0) - F_0(t_0)\big\}/\big\{\tfrac{3}{4}f_0(t_0)^2/h(t_0, t_0)\big\}^{1/3} > x\Big\}$$
$$\to I\!\!P\Big\{T(0) > x\big\{\tfrac{4}{3}f_0(t_0)h(t_0, t_0)\big\}^{-1/3}\Big\}, \; n \to \infty.$$

Now, $T(0)$ is the last time where

$$t \mapsto W\left(\tfrac{2}{3} h(t_0, t_0) f_0(t_0) t\right) + \tfrac{1}{3} h(t_0, t_0) f_0(t_0) t^2$$

reaches its minimum. By Brownian scaling, this means that

$$\tfrac{1}{2} T(0) \left\{ \tfrac{4}{3} f_0(t_0) h(t_0, t_0) \right\}^{1/3} = T(0) \left\{ \tfrac{1}{6} f_0(t_0) h(t_0, t_0) \right\}^{1/3}$$

is the last time where

$$t \mapsto W(t) + t^2, \; t \in \mathbb{R},$$

reaches its minimum. $\square$

As a final step in proving that the NPMLE $\hat{F}_n$ has the same limiting behavior as $F_n^{(1)}$, one would have to prove two things:

(1) $\hat{F}_n(t) - F_0(t_0) = \mathcal{O}_p(\delta_n)$, for t in an interval of the form $[t_0 - M\delta_n, t_0 + M\delta_n]$ (compare with Lemma 5.4).

(2) The process

$$\begin{aligned}
t \mapsto \; & \delta_n^{-2} \left\{ W_{\hat{F}_n}(t_0 + \delta_n t) - W_{\hat{F}_n} \right\} \\
& + \delta_n^{-2} \int_{[0, t_0 + \delta_n t]} \left(\hat{F}_n(t') - F_0(t_0) \right) dG_{\hat{F}_n}(t') \\
& - \delta_n^{-2} \int_{[0, t_0]} \left(\hat{F}_n(t') - F_0(t_0) \right) dG_{\hat{F}_n}(t')
\end{aligned} \tag{5.37}$$

has the same asymptotic behavior as the process $U_n^{(0)}$, defined in Lemma (5.5) (note that $U_n^{(0)}$ is (5.37), with $\hat{F}_n$ replaced by $F_n^{(1)}$). Assuming that the δ_n^{-1}–consistency of property 1 holds, we can, for $t \geq 0$, write (5.37) in the form

$$\begin{aligned}
U_n^{(0)}(t) - \delta_n^{-2} \int_{t_0 < t' \leq t_0 + \delta_n t, \, t' < x \leq u} & \frac{\hat{F}_n(u) - F_0(u)}{\left(F_0(u) - F_0(t') \right)^2} \, dP_n(x, t', u) \\
- \delta_n^{-2} \int_{t_0 < u \leq t_0 + \delta_n t, \, t' < x \leq u} & \frac{\hat{F}_n(t') - F_0(t')}{\left(F_0(u) - F_0(t') \right)^2} \, dP_n(x, t', u) + R_n(t),
\end{aligned}$$

where the process $t \mapsto R_n(t)$ will vanish in the limit, with a similar expansion for $t < 0$. The reason for believing that the other two processes added to $U_n^{(0)}(t)$ will also vanish in the limit, is that the integrands will have both positive and negative terms at the points t which have positive mass (since, roughly speaking, the expectation of $\hat{F}_n(t) - F_0(t)$ will be zero), and that $\hat{F}_n(T_{(i)}) - F_0(T_{(i)})$ and $\hat{F}_n(T_{(j)}) - F_0(T_{(j)})$ will (most likely) be almost independent for $T_{(i)}$ and $T_{(j)}$ far apart, (say $|T_{(i)} - T_{(j)}| > n^{-1/3} \log n$). The latter property is in contrast with the behavior of a term like

$$\delta_n^{-2} \int_{t_0 < t' \leq t_0 + \delta_n t, \, t' < x \leq u} \frac{\hat{F}_n(t') - F_0(t')}{\left(F_0(u) - F_0(t') \right)^2} \, dP_n(x, t', u),$$

where the restriction of t' to the interval $(t_0, t_0 + \delta_n t]$ causes strong dependence of terms of the above type.

The conjectures mentioned here have actually been supported by computer experiments, which also gave support of the conjecture that the asymptotic variance of the NPMLE at t_0 has the form

$$\left\{ \tfrac{3}{4} f_0(t_0)^2 / h(t_0, t_0) \right\}^{2/3} 4(EZ^2)\delta_n^2,$$

given in Theorem 5.3.

5.3 Deconvolution with a decreasing density

We consider the situation, discussed in Section 2.1. In this set-up we shall prove the following result for the (fictitious) 1-step estimator $F_n^{(1)}$.

Theorem 5.4. Let g be a right-continuous decreasing density on $[0, \infty)$, having only a finite number of discontinuity points $a_0 = 0 < a_1 < \ldots < a_m$. Moreover, suppose that g has a derivative $g'(x)$ at points $x \neq a_i$, $i = 0, \ldots, m$, satisfying

$$\int_{(0,\infty)} \frac{g'(x)^2}{g(x)}\, dx < \infty, \tag{5.38}$$

where the integrand is defined to be zero at the points a_i and at points x where g is zero, and where g' is bounded and continuous on the intervals (a_{i-1}, a_i), $i = 1, \ldots, m+1$, with $a_{m+1} \overset{\text{def}}{=} \infty$.

Furthermore, assume that there exist positive constants k_1 and k_2 such that the derivative g' of g satisfies the relation

$$|g'(t+u)| \leq k_1 |g'(t)|,$$

for all $t > 0$ and $0 < u < k_2$, such that $a_i < t < t+u < a_{i+1}$ for some i, $0 \leq i \leq m$.

Let the convolution density h be given by

$$h(z) = \int g(z-x)\, dF_0(x), \ z \geq 0,$$

where the distribution function F_0 of the (non-negative) random variables X_i, $1 \leq i \leq n$, is continuously differentiable at $z_0 > 0$, with derivative $f_0(z_0) > 0$ at z_0. Then

$$n^{1/3} \left\{ F_n^{(1)}(z_0) - F_0(z_0) \right\} f_0(z_0)^{-1/3} \left\{ 2 \sum_{i=0}^{m} \left(g(a_i) - g(a_i-) \right)^2 / h(z_0 + a_i) \right\}^{1/3} \overset{D}{\to} 2Z,$$

where $\overset{D}{\to}$ denotes convergence in distribution, and where Z is the last time where standard two-sided Brownian motion minus the parabola $y(t) = t^2$ reaches its maximum.

Example 5.1. If g is the uniform density on $[0, 1]$, then the two discontinuity points are $a_0 = 0$ and $a_1 = 1$. Since we define g to be right-continuous, we get $g(0) - g(0-) = 1$ and $g(1) - g(1-) = -1$. Let $F_0(1) = 1$, i.e., the probability distribution of the X_i's has support contained in $[0, 1]$. Theorem 5.4 now yields:

$$n^{1/3} \left\{ F_n^{(1)}(z_0) - F_0(z_0) \right\} \Big/ \left\{ \tfrac{1}{2} F_0(z_0)(1 - F_0(z_0)) f_0(z_0) \right\}^{1/3} \xrightarrow{\mathcal{D}} 2Z,$$

since

$$\sum_{i=0}^{m} (g(a_i) - g(a_i-))^2 / h(z_0 + a_i) = 1 / \{ F_0(z_0)(1 - F_0(z_0)) \}$$

in this case. It can also be deduced from Exercise 2 of Chapter 2 that in fact

$$n^{1/3} \left\{ \hat{F}_n(z_0) - F_0(z_0) \right\} \Big/ \left\{ \tfrac{1}{2} F_0(z_0)(1 - F_0(z_0)) f_0(z_0) \right\}^{1/3} \xrightarrow{\mathcal{D}} 2Z,$$

where $\hat{F}_n$ is the NPMLE of F_0. The latter result is also given in van Es (1991), Theorem 4.5, where it is shown that the variance of the NPMLE corresponds to a lower bound for the minimax risk of estimating F_0 at z_0, apart from a constant not depending on F_0 and the density g.

Example 5.2. If g is the exponential density on $[0, \infty)$, with scale parameter 1, Theorem 5.4 yields:

$$n^{1/3} \left\{ F_n^{(1)}(z_0) - F_0(z_0) \right\} \Big/ \left\{ \tfrac{1}{2} f_0(z_0) h(z_0) \right\}^{1/3} \xrightarrow{\mathcal{D}} 2Z.$$

According to our working hypothesis, we study the process $V_n^{(0)}$, defined by

$$V_n^{(0)}(t) = W_{F_0}(t) + \int_{[0,t]} \{ F_0(t') - F_0(t_0) \} \, dG_{F_0}(t'), \ t \geq 0,$$

For convenience of notation, we shall denote W_{F_0} by $W_n^{(0)}$ and W_{G_0} by $G_n^{(0)}$. The following limit result holds for the process $V_n^{(0)}$.

Lemma 5.7. Let, under the conditions of Theorem 5.4, the process $U_n^{(0)}$ be defined by

$$U_n^{(0)}(t) = n^{2/3} \{ V_n^{(0)}(z_0 + n^{-1/3}t) - V_n^{(0)}(z_0) \}, \ t \in \mathbb{R},$$

where $U_n^{(0)}(t) = 0$, if $t \leq -n^{1/3}t_0$. Then $U_n^{(0)}$ converges in distribution, in the topology of uniform convergence on compacta on the space of locally bounded real-valued functions on $\mathbb{R}$, to the process U, defined by

$$U(t) = \left\{ \sum_{i=0}^{m} (g(a_i) - g(a_i-))^2 / h(z_0 + a_i) \right\}^{1/2} W(t)$$
$$+ \tfrac{1}{2} \left\{ \sum_{i=0}^{m} (g(a_i) - g(a_i-))^2 / h(z_0 + a_i) \right\} f_0(z_0) t^2, \ t \in \mathbb{R},$$

where W is (standard) two-sided Brownian motion on $\mathbb{R}$, originating from zero.

Proof. We have, for $t > 0$:

$$n^{2/3}\{W_n^{(0)}(z_0 + tn^{-1/3}) - W_n^{(0)}(z_0)\} = n^{2/3} \sum_{z_0 < Z_{(i)} \le z_0 + tn^{-1/3}} \Delta W_n^{(0)}(Z_{(i)})$$

$$= n^{2/3} \int \frac{g(z - \tau_n^-) - g(z - \tau_n^+)}{\int g(z - x)\, dF_0(x)}\, dH_n(z),$$

(5.39)

where τ_n^- and τ_n^+ are defined by

$$\tau_n^- = \max\{Z_i : Z_i \le z_0\}, \text{ and } \tau_n^+ = \max\{Z_i : Z_i \le z_0 + tn^{-1/3}\}.$$

This follows from the fact that

$$\sum_{z_0 < Z_{(i)} \le z_0 + tn^{-1/3}} \int \frac{g(z - Z_{(i)}) - g(z - Z_{(i+1)})}{\int g(z - x)\, dF_0(x)}\, dH_n(z)$$

$$= \int \frac{g(z - \tau_n^-) - g(z - \tau_n^+)}{\int g(z - x)\, dF_0(x)}\, dH_n(z),$$

(telescoping sums). It now follows from (5.38) and the representation (5.39) that the variance of

$$n^{2/3}\{W_n^{(0)}(z_0 + tn^{-1/3}) - W_n^{(0)}(z_0)\}$$

(5.40)

converges to

$$t \cdot \sum_{i=0}^{m} (g(a_i) - g(a_i-))^2 / h(z_0 + a_i),$$

since the remaining part of the variance of (5.40) is bounded above by

$$c \cdot n^{-1/3} \int \frac{g'(x)^2}{g(x)}\, dx.$$

Here we use that, for some $\delta > 0$,

$$\int \frac{(g(z - z_0) - g(z - z_0 - t \cdot n^{-1/3}))^2}{h(z)}\, dz$$

$$\le \delta^{-1} \int \frac{(g(z - z_0) - g(z - z_0 - t \cdot n^{-1/3}))^2}{g(z - z_0 - t \cdot n^{-1/3})}\, dz$$

which follows from the fact that the density f_0 satisfies $f_0(z) \ge \epsilon > 0$, for some $\epsilon > 0$, in a neighborhood $[z_0, z_0 + \eta]$ of z_0, implying

$$h(z) \ge \tfrac{1}{2} g(z - z_0 - t \cdot n^{-1/3})\, \epsilon \eta,$$

for large n. We also use the boundedness and continuity of g' on intervals between points of jump of g and apply the mean value theorem.

Next we note that, under the conditions of the Lemma,

$$\int \frac{g(z - z_0) - g(z - z_0 - tn^{-1/3})}{\int g(z - x)\, dF_0(x)}\, dH(z) = 0$$

and by empirical process theory it is easily seen that the process

$$t \mapsto n^{2/3} \int \frac{g(z - z_0) - g(z - z_0 - tn^{-1/3})}{\int g(z - x)\, dF_0(x)}\, d(H_n - H)(z)$$

converges in distribution, in the topology of uniform convergence on compacta on the space of locally bounded real-valued functions on $\mathbb{R}$, to the process

$$t \mapsto \left\{ \sum_{i=0}^{m} (g(a_i) - g(a_i-))^2 / h(z_0 + a_i) \right\}^{1/2} W(t),\ t \geq 0.$$

Since $\tau_n^- - z_0$ and $\tau_n^+ - z_0 - tn^{-1/3}$ are $\mathcal{O}_p(n^{-1}\log n)$, uniformly in t, the process (5.39) has the same limiting behavior.

Similarly, defining $G_n^{(0)} = G_{F_0}$, we get that the process

$$n^{2/3} \int_{(z_0, z_0 + n^{-1/3}t]} \left\{ F_0(t') - F_0(t_0) \right\} dG_n^{(0)}(t'),\ t \geq 0,$$

converges on bounded intervals almost surely to the deterministic function

$$t \mapsto \tfrac{1}{2} \left\{ \sum_{i=0}^{m} (g(a_i) - g(a_i-))^2 / h(z_0 + a_i) \right\} f_0(z_0) t^2,\ t \geq 0.$$

To see this, note that, by (2.12), we have to deal with terms of the form

$$\{ F_0(Z_{(i+1)}) - F_0(z_0) \} \Delta G_F(Z_{(i)})$$
$$= \{ F_0(Z_{(i+1)}) - F_0(z_0) \} \int \frac{\left\{ g(z - Z_{(i)}) - g(z - Z_{(i+1)}) \right\}^2}{\left\{ \int g(z - x)\, dF_0(x) \right\}^2}\, dH_n(z),$$

and $\left\{ g(z - Z_{(i)}) - g(z - Z_{(i+1)}) \right\}^2$ behaves as $\left\{ g'(z - Z_{(i)})(Z_{(i+1)}) - Z_{(i)}) \right\}^2$ at the interior of the set of points $z - Z_{(i)}$ where g' is differentiable.

Since negative values of t can be dealt with in a completely similar way, the result follows. $\square$

In accordance with the methods used in the preceding sections, we introduce, for each $a > 0$, the random variable $T_n^{(0)}(a)$ by

$$T_n^{(0)}(a) = \sup\{ t \geq 0 : V_n^{(0)}(t) - (a - F_0(t_0)) G_n^{(0)}(t) \text{ is minimal} \}.$$

Likewise, in analogy with (5.9) and (5.35), we define the process $\{T(a) : a \in \mathbb{R}\}$ by

$$T(a) = \sup\left\{t \in \mathbb{R} : U(t) - a \cdot \left\{\sum_{i=0}^{m}(g(a_i) - g(a_i-))^2/h(z_0 + a_i)\right\}t \text{ is minimal}\right\}.$$

As in Lemma 5.2, we get that the process

$$\left\{n^{1/3}\{T_n^{(0)}(a_0 + n^{-1/3}a) - t_0\} : a \in \mathbb{R}\right\}$$

converges in distribution to the process $\{T(a) : a \in \mathbb{R}\}$. This follows from the following lemma.

Lemma 5.8. For each $\epsilon > 0$ and $M_1 > 0$ an $M_2 > 0$ can be found such that

$$\mathbb{P}\left\{\max_{a \in [-M_1, M_1]} n^{1/3}\{T_n^{(0)}(a_0 + n^{-1/3}a) - t_0\} > M_2\right\} < \epsilon,$$

and

$$\mathbb{P}\left\{\min_{a \in [-M_1, M_1]} n^{1/3}\{T_n^{(0)}(a_0 + n^{-1/3}a) - t_0\} < -M_2\right\} < \epsilon,$$

for all sufficiently large n.

Proof. The proof proceeds along similar lines as the proof of Lemma 5.3.

Let $Z_n^{(0)}(u) = \{W_n^{(0)}(z_0 + u) - W_n^{(0)}(z_0)\}^2$. Then,

$$\mathbb{P}\left\{\sup_{0 \leq u \leq u_0} Z_n^{(0)}(u)\right\} \leq cn^{-1}\left\{u_0 \cdot \sum_{i=0}^{m}(g(a_i) - g(a_i-))^2/h(z_0 + a_i) + \mathcal{O}(u_0^2)\right\},$$

for a constant $c > 0$, if u_0 satisfies $H(t_0 + u_0) < 1$. This can, e.g., be seen by applying Lemma 4.1 in Kim and Pollard (1990) to the functions ϕ_β, defined by

$$\phi_\beta(u, \beta) = (g(u - z_0) - g(u - z_0 - \beta))/h(u), \ \beta \geq 0, \ u \geq z_0.$$

Under the conditions of Theorem 5.4 the class of functions $\{|\phi_\beta| : 0 \leq \beta \leq R\}$ has an envelope

$$G_R(u) = h(u)^{-1}\left\{\sum_{i=0}^{m}|g(a_i) - g(a_i-)| \ 1_{[z_0+a_i, z_0+a_i+R]}(u)\right.$$
$$\left. + R\sup_{t \in [0,R], \, t \notin \{a_0,...,a_m\}}|g'(t)|\right\},$$

for $u \geq z_0$, and we get

$$PG_R^2 \leq c_1 R + c_2 \cdot R^2,$$

for positive constants c_1, c_2. The remaining part of the argument is similar to the proof of Lemma 5.3.   □

We can now give the proof of Theorem 5.4.

Proof of Theorem 5.4. First of all, we have:

$$\mathbb{P}\{F_n^{(1)}(t_0) - F_0(t_0) > n^{-1/3}a\} = \mathbb{P}\{T_n^{(0)}(a_0 + n^{-1/3}a) < t_0\},$$

where $a_0 = F_0(t_0)$. Using Lemmas 5.7 and 5.8, we get that the process

$$\left\{n^{1/3}\{T_n^{(0)}(a_0 + n^{-1/3}a) - t_0\} : a \in \mathbb{R}\right\}$$

converges in the Skorohod topology on $D(\mathbb{R})$ to the process $\{T(a) : a \in \mathbb{R}\}$. As in the preceding sections, the process

$$\{T(a) - f_0(t_0)^{-1}a : a \in \mathbb{R}\}$$

is stationary. Hence,

$$\mathbb{P}\left\{n^{1/3}\{F_n^{(1)}(t_0) - F_0(t_0)\}\left\{2\sum_{i=0}^{m}\frac{(g(a_i) - g(a_i-))^2}{(f_0(z_0)h(z_0 + a_i))}\right\}^{1/3} > x\right\}$$

$$\rightarrow \mathbb{P}\left\{T(0) > x\Big/\left\{2f_0(z_0)^2\sum_{i=0}^{m}\frac{(g(a_i) - g(a_i-))^2}{h(z_0 + a_i)}\right\}^{1/3}\right\}, \quad n \rightarrow \infty.$$

Since $T(0)$ is the last time where

$$t \mapsto W\left(t\sum_{i=0}^{m}(g(a_i) - g(a_i-))^2/h(z_0 + a_i)\right) + \tfrac{1}{2}f_0(z_0)t^2\sum_{i=0}^{m}\frac{(g(a_i) - g(a_i-))^2}{h(z_0 + a_i)}$$

reaches its minimum, this means that

$$\tfrac{1}{2}T(0)\left\{2f_0(z_0)^2\sum_{i=0}^{m}(g(a_i) - g(a_i-))^2/h(z_0 + a_i)\right\}^{1/3}$$

is the last time where

$$t \mapsto W(t) + t^2, \quad t \in \mathbb{R},$$

reaches its minimum. □

Computer experiments, using simulated samples of convolution densities, again indicate that in many cases the NPMLE has the same asymptotic behavior as the 1-step estimator $F_n^{(1)}$. But for a proof of this, one has to prove two final steps, which are of the same type as the two remaining steps in the proof for the asymptotic behavior of the NPMLE for interval censoring, Case 2.

We will deal with this problem elsewhere. It seems that the limiting behavior of the NPMLE is the same as that of the 1-step estimator, given in Theorem 5.4, at least when the support of F_0 is contained in an interval not bigger than the minimum distance between discontinuities of g. If the latter condition is not fulfilled, it is not clear that the limiting behavior will be the same.

5.4   Estimation of the mean

As can be expected from the general theory on differentiable functionals (see e.g., van der Vaart (1991) and part I, section 3), efficient estimators of smooth functionals like the mean

$$\mu_{F_0} = \int t \, dF_0(t)$$

should have $\sqrt{n}$−behavior. We will discuss a result of this type for interval censoring, Case 1, using the same set-up as in section 4.1. Let $(X_1, T_1), \ldots, (X_n, T_n)$ be a sample of random variables in $\mathbb{R}^2_+$, where X_i and T_i are independent (non-negative) random variables with continuous distribution functions F_0 and G, respectively.

Furthermore, we assume that the support of P_{F_0} is a bounded interval $I = [0, M]$, and that F_0 and G have densities f_0 and g, respectively, satisfying

$$g(t) \geq \delta > 0, \text{ and } f_0(t) \geq \delta > 0, \quad \text{if } t \in I,$$

for some $\delta > 0$. Finally we assume that g has a bounded derivative on I. An example of this situation is the case where F_0 and G are both the uniform distribution function on $[0, 1]$. It is certainly possible to prove the following theorem under weaker conditions, but at the cost of an increasing number of technicalities.

Theorem 5.5. Let F_0 and G satisfy the conditions, listed above, and let $\hat{F}_n$ be the NPMLE of F_0. Then

$$\sqrt{n} \int_I \left(\hat{F}_n(t) - F_0(t) \right) dt \xrightarrow{\mathcal{D}} U,$$

where U has a normal distribution with mean zero and variance

$$\sigma^2_{F_0} = \int \frac{F_0(t)\left(1 - F_0(t)\right)}{g(t)} \, dt.$$

Before starting the proof of Theorem 5.5, we give an outline of the steps in the argument. First we note that

$$\int \frac{\{x > t\}}{g(t)} \, dP(x, t) = \int_I \left(1 - F_0(t)\right) dt = \mu_{F_0},$$

where P is the probability measure of a pair (X_i, T_i). Secondly, let $\tau_1 < \tau_2 < \ldots < \tau_m$ be the points of jump of $\hat{F}_n$ on the interval $[0, T_{(n)}]$, and let $\tau_0 = 0$, $\tau_{m+1} = T_{(n)}$. Then, defining the function $\tilde{g}_n$ on I by $\tilde{g}_n(0) = g(0)$, and

$$\tilde{g}_n(t) = g(\tau_i), \ \tau_{i-1} < t \leq \tau_i, \ 1 \leq i \leq m + 1,$$

we get

$$
\int_I (1 - \hat{F}_n(t))\, dt - \mu_{F_0} = \int \frac{1 - \hat{F}_n(t) - \{x > t\}}{g(t)}\, dP(x,t)
$$

$$
= \int \frac{1 - \hat{F}_n(t) - \{x > t\}}{\tilde{g}_n(t)}\, dP(x,t)
$$

$$
+ \int \{1 - \hat{F}_n(t) - \{x > t\}\} \left\{ \frac{1}{g(t)} - \frac{1}{\tilde{g}_n(t)} \right\} dP(x,t), \qquad (5.41)
$$

$$
= \int \frac{1 - \hat{F}_n(t) - \{x > t\}}{\tilde{g}_n(t)}\, dP(x,t)
$$

$$
- \int \{\hat{F}_n(t) - F_0(t)\} \left\{ \frac{1}{g(t)} - \frac{1}{\tilde{g}_n(t)} \right\} dG(t).
$$

By the characterization of the NPMLE, given in Proposition 1.2, we have:

$$
\int \frac{1 - \hat{F}_n(t) - \{x > t\}}{\tilde{g}_n(t)}\, dP_n(x,t) = 0,
$$

where P_n is the empirical measure of $(X_1, T_1), \ldots, (X_n, T_n)$. Hence

$$
\sqrt{n} \int \frac{1 - \hat{F}_n(t) - \{x > t\}}{\tilde{g}_n(t)}\, dP(x,t)
$$
$$
= -\sqrt{n} \int \frac{1 - \hat{F}_n(t) - \{x > t\}}{\tilde{g}_n(t)}\, d(P_n - P)(x,t). \qquad (5.42)
$$

We shall prove that the right-hand side of (5.42) converges to U, where U is defined as in Theorem 5.5, and that the last term of (5.41) is $o_P(n^{-1/2})$.

The distance between the successive jumps of $\hat{F}_n$ plays an essential role in the proof. Roughly speaking, they are of order $\mathcal{O}_p(n^{-1/3})$, but we need a uniform upper bound for this distance. To this end we introduce, for $a \in (0, 1)$, the random variables $T_n(a)$, defined by

$$
T_n(a) = \sup\{t \in [0, T_{(n)}] : V_n(t) - aG_n(t) \text{ is minimal}\}, \qquad (5.43)
$$

where G_n is the empirical distribution function of $T_1, \ldots, T_n$, and V_n is defined by

$$
V_n(t) = n^{-1} \sum_{T_i \leq t} 1_{\{X_i \leq T_i\}} = \int_{u \leq t} 1_{\{x \leq u\}}\, dP_n(x, u). \qquad (5.44)
$$

Note the similarity with the definitions of the random variables $T_n^{(0)}(a)$, discussed in the preceding sections.

The process $\left\{ (G_n(T_n(a)), V_n(T_n(a))) : a \in (0, 1) \right\}$ runs through the vertices of the cumulative sum diagram, consisting of the point $(0, 0)$ and the points

$$
(G_n(T_{(i)}), V_n(T_{(i)})), \ 1 \leq i \leq n.
$$

The following lemma gives an upper bound for the probability that the distance of successive jumps of $\hat{F}_n$ is bigger that $n^{-1/3} \log n$. The proof is quite similar to the proof of Lemma 2.4 in Groeneboom (1987). It is based on a stopping time argument for a suitably chosen exponential martingale.

Lemma 5.9. Let, for each $a \in (0,1)$, $F_0^{-1}(a)$ be defined by $F_0(F_0^{-1}(a)) = a$ (note that $F_0^{-1}(a)$ is well-defined, since F_0 has a strictly positive density on I). Then, for each $a \in (0,1)$,

$$\mathbb{P}\{|T_n(a) - F_0^{-1}(a)| \geq n^{-1/3} \log n\} \leq c_1 \exp\{-c_2 (\log n)^2\},$$

for positive constants c_1 and c_2, not depending on a and n.

Proof. Let $a \in (0,1)$ and suppose $t_0 = F_0^{-1}(a)$. Let the process U_n be defined by

$$U_n(t) = V_n(t) - V_n(t_0) - a\{G_n(t) - G_n(t_0)\}, \ t \in I.$$

Then we have

$$\mathbb{P}\{|T_n(a) - F_0^{-1}(a)| \geq n^{-1/3} \log n\}$$
$$\leq \mathbb{P}\{U_n(t) \leq 0, \ \text{for some} \ t \in I \backslash (t_0 - n^{-1/3} \log n, t_0 + n^{-1/3} \log n)\}.$$

We only give an upper bound for the probability

$$\mathbb{P}\{U_n(t) \leq 0, \ \text{for some} \ t \geq t_0 + n^{-1/3} \log n\},$$

since the probability

$$\mathbb{P}\{U_n(t) \leq 0, \ \text{for some} \ t \leq t_0 - n^{-1/3} \log n\},$$

can be treated similarly.

Define the process W_n by

$$W_n(t) = U_n(t) - \int_{[t_0, t]} \{F_0(t') - F_0(t_0)\} \, dG_n(t'), \ t \in I, \ t \geq t_0.$$

Moreover, we define $\Delta W_{n,i}$ and $\Delta G_{n,i}$ by

$$\Delta W_{n,i} = W_n(T_{(i)}) - W_n(T_{(i-1)}), \ \text{and} \ \Delta G_{n,i} = G_n(T_{(i)}) - G_n(T_{(i-1)}),$$

where $T_{(0)} \overset{\text{def}}{=} 0$. Then we can write

$$\mathbb{P}\{U_n(t) \leq 0, \ \text{for some} \ t \geq t_0 + n^{-1/3} \log n\}$$
$$= \mathbb{P}\Big\{W_n(t) + \int_{[t_0, t]} \{F_0(t') - F_0(t_0)\} \, dG_n(t') \leq 0,$$
$$\text{for some} \ t \geq t_0 + n^{-1/3} \log n\Big\}.$$

Conditionally on the observation times T_i, the terms $\Delta W_{n,i}$, with $T_{(i)} \in [t_0, t]$ are independent random variables, with values in $[-n^{-1}, n^{-1}]$, and we have, for $t \geq t_0$,

$$
E\left\{\exp\{-n^{2/3} W_n(t)\} \mid T_1, T_2, \ldots\right\}
$$

$$
= E\left\{\exp\left\{-n^{2/3} \sum_{T_{(i)} \in [t_0, t]} \Delta W_{n,i}\right\} \mid T_1, T_2, \ldots\right\}
$$

$$
= \prod_{T_{(i)} \in [t_0, t]} \left\{F_0(T_{(i)}) \exp\{-n^{-1/3}(1 - F_0(T_{(i)}))\}\right.
$$
$$
\left. + (1 - F_0(T_{(i)})) \exp\{n^{-1/3} F_0(T_{(i)})\}\right\}
$$
$$
= \exp\left\{\tfrac{1}{2}(1 + r_n(t)) n^{-2/3} \sum_{T_{(i)} \in [t_0, t]} F_0(T_{(i)})(1 - F_0(T_{(i)}))\right\}
$$
$$
= \exp\left\{\tfrac{1}{2}(1 + r_n(t)) n^{1/3} \int_{[t_0, t]} F_0(t')(1 - F_0(t'))\, dG_n(t')\right\}
$$

(5.45)

where $r_n(t)$ tends uniformly to zero for each realization of the observation times $T_{(i)}$, since $n^{-1/3} F_0(T_{(i)})$ and $n^{-1/3}(1 - F_0(T_{(i)}))$ tend uniformly to zero, as $n \to \infty$.

Let the process $\{Z_n(t) : t \in I, t \geq t_0\}$ be defined by

$$
Z_n(t) = \exp\{-n^{2/3} W_n(t)\} \,/\, E\left\{\exp\{-n^{2/3} W_n(t)\} \mid T_1, T_2, \ldots\right\}.
$$

Then, conditionally on the observation times $T_1, \ldots, T_n$, the process Z_n is a martingale with respect to the filtration $\{\mathcal{F}_{n,t} : t \in I, t \geq t_0\}$, where

$$
\mathcal{F}_{n,t} = \sigma\left\{1_{\{X_i \leq T_i\}} : T_i \in [t_0, t]\right\}.
$$

We now define the stopping time τ_n by

$$
\tau_n = \inf\{t \in [t_0 + n^{-1/3} \log n, M] : U_n(t) \leq 0\}.
$$

If $U_n(t) > 0$, for all $t \in [t_0 + n^{-1/3} \log n, M]$, we define $\tau_n = \infty$. We shall derive an upper bound for the probability

$$
\mathbb{P}\{\tau_n < \infty \mid T_1, \ldots, T_n\},
$$

and denote the conditional probabilities $\mathbb{P}\{\cdot \mid T_1, \ldots, T_n\}$ and the conditional expectations $E\{\cdot \mid T_1, \ldots, T_n\}$ both by $\mathbb{P}_n$.

Since (in the conditional set-up), Z_n is a martingale, we have

$$
\mathbb{P}_n Z_n(M \wedge \tau_n) = \mathbb{P}_n Z_n(t_0) = 1.
$$

Furthermore, by (5.45) we have:

$$
\mathbb{P}_n Z_n(M \wedge \tau_n) \geq \mathbb{P}_n Z_n(\tau_n) 1_{\{\tau_n < \infty\}}
$$
$$
= \mathbb{P}_n \exp\left\{-n^{2/3} W_n(\tau_n) - \tfrac{1}{2}(1 + r_n(\tau_n)) n^{1/3}\right.
$$
$$
\left. \int_{[t_0, \tau_n]} F_0(t)(1 - F_0(t))\, dG_n(t)\right\} 1_{\{\tau_n < \infty\}}.
$$

(5.46)

But for $t \geq t_0 + n^{-1/3} \log n$ we have, if $U_n(t) \leq 0$,

$$-n^{2/3} W_n(t) - \tfrac{1}{2}(1 + r_n(t)) n^{1/3} \int_{[t_0,t]} F_0(t')(1 - F_0(t')) \, dG_n(t')$$

$$= -n^{2/3} U_n(t) + n^{2/3} \int_{[t_0,t]} \{F_0(t') - F_0(t_0)\} \, dG_n(t')$$

$$- \tfrac{1}{2}(1 + r_n(t)) n^{1/3} \int_{[t_0,t]} F_0(t')(1 - F_0(t')) \, dG_n(t')$$

$$\geq n^{2/3} \int_{[t_0,t]} \{F_0(t') - F_0(t_0)\} \, dG_n(t')$$

$$- \tfrac{1}{2}(1 + r_n(t)) n^{1/3} \int_{[t_0,t]} F_0(t')(1 - F_0(t')) \, dG_n(t')$$

$$\geq \tfrac{1}{2} f_0(t_0) n^{2/3} \int_{[t_0,t_0+n^{-1/3}\log n]} (t' - t_0) \, dG_n(t')$$
$$+ \tfrac{1}{2} f_0(t_0) n^{1/3} \{G_n(t) - G_n(t_0 + n^{-1/3} \log n)\} \log n$$
$$- \tfrac{1}{2}(1 + r_n(t)) n^{1/3} \int_{[t_0,t]} F_0(t')(1 - F_0(t')) \, dG_n(t')$$

$$\geq \tfrac{1}{2} f_0(t_0) n^{2/3} \int_{[t_0,t_0+n^{-1/3}\log n]} \left(\frac{t - t_0 - 2n^{-1/3} F_0(t)(1 - F_0(t))}{f_0(t_0)} \right) dG_n(t),$$

$$(5.47)$$

for $n \geq n_0$, where n_0 only depends on $f_0(t_0)$ (and not on t or G_n). Here we use that, by the conditions on F_0, $F_0(t) - F_0(t_0) \geq \tfrac{1}{2} f_0(t_0)(t - t_0)$, if $0 \leq t - t_0 \leq n^{-1/3} \log n$ and n is sufficiently large. From (5.46) and (5.47) we obtain, for $n \geq n_0$:

$$\mathbb{P}_n\{\tau_n < \infty\}$$
$$\leq \mathbb{P}_n \exp\Big\{ -\tfrac{1}{2} n^{2/3} f_0(t_0)$$
$$\int_{[t_0,t_0+n^{-1/3}\log n]} \left(t - t_0 - 2n^{-1/3} \frac{F_0(t)(1 - F_0(t))}{f_0(t_0)} \right) dG_n(t) \Big\}.$$

$$(5.48)$$

By (5.48), the probability of a large deviation of the process U_n is reduced to the probability of a large deviation of the empirical distribution function G_n. We now note that

$$n^{2/3} \int_{[t_0,t_0+n^{-1/3}\log n]} \left(t - t_0 - 2n^{-1/3} F_0(t)(1 - F_0(t))/f_0(t_0) \right) dG_n(t)$$

$$= n^{2/3} \int_{[t_0,t_0+n^{-1/3}\log n]} \left(t - t_0 - 2n^{-1/3} F_0(t)(1 - F_0(t))/f_0(t_0) \right) dG(t)$$

$$+ n^{2/3} \int_{[t_0,t_0+n^{-1/3}\log n]} \left(\frac{t - t_0 - 2n^{-1/3} F_0(t)(1 - F_0(t))}{f_0(t_0)} \right) d(G_n - G)(t)$$

$$= (1 + o(1)) g(t_0) \Big\{ \tfrac{1}{2}(\log n)^2 - \frac{2F_0(t_0)(1 - F_0(t_0))}{f_0(t_0)} \log n \Big\}$$

$$+ n^{2/3} \int_{[t_0,t_0+n^{-1/3}\log n]} \left(\frac{t - t_0 - 2n^{-1/3} F_0(t)(1 - F_0(t))}{f_0(t_0)} \right) d(G_n - G)(t),$$

and, moreover, that

$$\text{var}\left\{ n^{2/3} \int_{[t_0, t_0+n^{-1/3}\log n]} \left(\frac{t \leftarrow t_0 - 2n^{-1/3}F_0(t)\big(1-F_0(t)\big)}{f_0(t_0)} \right) d(G_n - G)(t) \right\}$$

$$\sim n^{1/3}g(t_0) \int_{[t_0, t_0+n^{-1/3}\log n]} \left(\frac{t - t_0 - 2n^{-1/3}F_0(t)\big(1-F_0(t)\big)}{f_0(t_0)} \right)^2 dt$$

$$\sim \tfrac{1}{3} n^{-2/3} g(t_0)(\log n)^3, \text{ as } n \to \infty.$$

Hence, by Bernstein's inequality (see, e.g., Pollard (1984), p. 193), we get:

$$\mathbb{P}\left\{ n^{2/3} \int_{[t_0, t_0+n^{-1/3}\log n]} \left(\frac{t - t_0 - 2n^{-1/3}F_0(t)\big(1-F_0(t)\big)}{f_0(t_0)} \right) d(G_n - G)(t) \right.$$
$$\left. \leq -\tfrac{1}{2}g(t_0)\left\{ \tfrac{1}{2}(\log n)^2 - \frac{2F_0(t_0)\big(1-F_0(t_0)\big)}{f_0(t_0)} \log n \right\} \right\}$$
$$\leq 2\exp\{-c \cdot n^{2/3}/\log n\},$$

for a constant $c > 0$. Applying this on (5.48), we obtain

$$\mathbb{P}\{\tau_n < \infty\} \leq \exp\left\{ -\tfrac{1}{4}f_0(t_0)g(t_0)\left\{ \tfrac{1}{2}(\log n)^2 - \frac{2F_0(t_0)\big(1-F_0(t_0)\big)}{f_0(t_0)} \log n \right\} \right\}$$
$$+ 2\exp\{-c \cdot n^{2/3}/\log n\}.$$

This gives the desired bound, since

$$\mathbb{P}\{U_n(t) \leq 0, \text{ for some } t \geq t_0 + n^{-1/3}\log n\} = \mathbb{P}\{\tau_n < \infty\},$$

and since $f_0(t)$ and $g(t)$ are uniformly bounded away from zero for $t \in I$.   □

Now let $a_1, a_2, \ldots$ be such that

$$a_i = i \cdot n^{-1/3}\log n, \ i = 1, 2, \ldots,$$

and let $m_n = [n^{1/3}/\log n]$, where $[a]$ denotes the biggest integer $\leq a$. Then we get from Lemma 5.9:

$$\mathbb{P}\left\{ |T_n(a_i) - F_0^{-1}(a_i)| \geq n^{-1/3}\log n \right\} \leq c_1 \exp\{-c_2(\log n)^2\}, \ 1 \leq i \leq m_n.$$

This implies, by the monotonicity of the function $a \mapsto T_n(a)$ and the conditions on F_0 that there exists a constant $c_3 > 0$ such that

$$\mathbb{P}\left\{ \sup_{a \in (0,1)} |T_n(a) - F_0^{-1}(a)| \geq c_3 n^{-1/3}\log n \right\} \leq m_n c_1 \exp\{-c_2(\log n)^2\}$$
$$\leq c_1 \exp\{-\tfrac{1}{2}c_2(\log n)^2\},$$

for sufficiently large n. This, in turn, implies that both the maximum distance between two successive points of jump of $\hat{F}_n$ and the maximum distance between $\hat{F}_n$ and F_0 are of order $n^{-1/3} \log n$. Hence we get, for the last term in (5.41):

$$\int \{\hat{F}_n(t) - F_0(t)\} \left\{ \frac{1}{g(t)} - \frac{1}{\tilde{g}_n(t)} \right\} dG(t) = \mathcal{O}_p(n^{-2/3}(\log n)^2),$$

where we use that g has a bounded derivative on I. Furthermore, by (5.42):

$$\sqrt{n} \int \frac{1 - \hat{F}_n(t) - \{x > t\}}{\tilde{g}_n(t)} \, dP(x, t)$$

$$= -\sqrt{n} \int \frac{1 - \hat{F}_n(t) - \{x > t\}}{\tilde{g}_n(t)} \, d(P_n - P)(x, t)$$

$$= -\sqrt{n} \int \frac{1 - F_0(t) - \{x > t\}}{g(t)} \, d(P_n - P)(x, t)$$

$$\quad - \sqrt{n} \int \{1 - F_0(t) - \{x > t\}\} \left\{ \frac{1}{\tilde{g}_n(t)} - \frac{1}{g(t)} \right\} d(P_n - P)(x, t)$$

$$\quad - \sqrt{n} \int \frac{\hat{F}_n(t) - F_0(t)}{\tilde{g}_n(t)} \, d(P_n - P)(x, t)$$

Since the class of functions

$$\{(F - F_0)/g_n : \quad F \text{ is a distribution function, and } g_n \text{ is a piecewise}$$
$$\text{constant version of } g, \text{ with distance } \leq c \cdot n^{-1/3} \log n$$
$$\text{between successive points of jump}\}$$

is "manageable" in the sense of Pollard (1989) (see Birman and Solomjak (1967) and van de Geer (1990) for the relevant entropy calculations), it follows that

$$\sqrt{n} \int \frac{\hat{F}_n(t) - F_0(t)}{\tilde{g}(t)} \, d(P_n - P)(x, t) = o_p(1), \ n \to \infty.$$

Similarly,

$$\sqrt{n} \int \{1 - F_0(t) - \{x > t\}\} \left\{ \frac{1}{\tilde{g}(t)} - \frac{1}{g(t)} \right\} d(P_n - P)(x, t) = o_p(1).$$

Theorem 5.5 now follows from (5.41).

5.5 Exercises

1. Let $T_n(a)$ be defined by (5.43), and let the conditions of Theorem 5.1 be satisfied. Moreover, let $a_0 = F_0(t_0)$, where t_0 satisfies the conditions of Theorem 5.1. Show, using the same techniques as in the proof of Lemma 5.3, that for each $M_1 > 0$ and $\epsilon > 0$ there exists an $M_2 > 0$ such that

$$\mathbb{P}\left\{ \max_{x \in [-M_1, M_1]} n^{1/3} |T_n(a_0 + xn^{-1/3}) - t_0| \geq M_2 \right\} < \epsilon,$$

for all sufficiently large n.

2. Let $U_n(t)$ be defined as in the proof of Lemma 5.9. Show that the process

$$t \mapsto \begin{cases} n^{2/3} U_n(t_0 + n^{-1/3}t) & , t_0 + n^{-1/3}t \in I, \\ 0 & , t_0 + n^{-1/3}t \notin I, \end{cases}$$

converges in distribution, in the topology of uniform convergence on compacta, to the process U, defined by

$$U(t) = \sqrt{g(t_0)F_0(t_0)\big(1 - F_0(t_0)\big)}\, W(t) + \tfrac{1}{2}f_0(t_0)g(t_0)t^2, \ t \in \mathbb{R},$$

where W is standard two-sided Brownian motion on $\mathbb{R}$, originating from zero.

3. Let $T_n(a)$ and $U(t)$ be defined as in Exercises 1 and 2. Show that the process

$$\left\{ n^{1/3}\big\{ T_n(a_0 + n^{-1/3}a) - t_0 \big\} : a \in \mathbb{R} \right\}$$

converges in distribution, in the space $D(\mathbb{R})$ with the Skorohod topology, to the process $\{T(a) : a \in \mathbb{R}\}$, where

$$T(a) = \sup\big\{ t \in \mathbb{R} : U(t) - a \cdot tg(t_0) \ \text{is minimal} \big\}.$$

4. Deduce Theorem 5.1 from Exercises 1 to 3.

5. Let, in Theorem 5.4, g be the standard exponential density on $[0, \infty)$, and let F_0 and z_0 be such that the conditions of Theorem 5.4 are satisfied. Deduce from Exercise 3 in Chapter 2:

$$n^{1/3}\big\{ \hat{F}_n(z_0) - F_0(z_0) \big\} / \big\{ \tfrac{1}{2}f_0(z_0)h(z_0) \big\}^{1/3} \xrightarrow{\mathcal{D}} 2U,$$

where U is the last time where standard two-sided Brownian motion minus the parabola $y(t) = t^2$ reaches its maximum, and $\hat{F}_n$ is the NPMLE of F_0.

6. Deduce from Exercise 2 in Chapter 2 that, under the same conditions and with the same notation as in Exercise 5, but with g replaced by the uniform density on $[0, 1]$,

$$n^{1/3}\big\{ \hat{F}_n(z_0) - F_0(z_0) \big\} / \big\{ \tfrac{1}{2}F_0(z_0)\big(1 - F_0(z_0)\big)f_0(z_0) \big\}^{1/3} \xrightarrow{\mathcal{D}} 2U.$$

7. Let F_0 and G satisfy the same conditions as in Theorem 5.5, and let $\hat{F}_n$ be the NPMLE of F_0. Show that

$$\sqrt{n}\left\{ \int_I \hat{F}_n^2(t)\, dt - \int_I F_0^2(t)\, dt \right\} \xrightarrow{\mathcal{D}} U,$$

where U has a normal distribution with mean zero and variance

$$\sigma_{F_0}^2 = 4 \int \frac{F_0^3(t)\big(1 - F_0(t)\big)}{g(t)}\, dt.$$

References

[1] Ayer, M., Brunk, H.D., Ewing, G.M., Reid, W.T., Silverman, E. (1955), *An empirical distribution function for sampling with incomplete information*, Ann. Math. Statist. **26**, 641–647.

[2] Barlow, R.E., Bartholomew, D.J., Bremner, J.M., Brunk, H.D. (1972), *Statistical Inference under Order Restrictions*, Wiley, New York.

[3] Begun, J. M., Hall, W. J., Huang, W. M., and Wellner, J. A. (1983), *Information and asymptotic efficiency in parametric-nonparametric models* , Ann. Statist. **11**, 432–452.

[4] Bennett, G., (1962), *Probability inequalities for sums of independent random variables*, J. Amer. Statist. Assoc. **57**, 33–45.

[5] Beran, R. (1977a), *Estimating a distribution function*, Ann. Statist. **5**, 400–404.

[6] Beran, R. (1977b), *Robust location estimates*, Ann. Statist. **5**, 431–444.

[7] Bickel, P. J., Klaassen, C. A. J., Ritov, Y., and Wellner, J. A. (1992), *Efficient and Adaptive Estimation for Semiparametric Models*, Forthcoming monograph, Johns Hopkins University Press, Baltimore.

[8] Birgé, L., Massart, P., (1991), *Rates of convergence for minimum contrast estimators*, Rapport Technique Nr. 140, Université Paris VI, Laboratoire de Statistique Théorique et Appliquée.

[9] Birman, M.Š., Solomjak, M.Z., (1967), *Piecewise-polynomial approximations of functions in the classes W_p^α*, Math. Sbornik. **73**, 295–317.

[10] Carroll, R.J., Hall, P., (1988), *Optimal rates of convergence for deconvolving a density*, J. Amer. Statist. Assoc. **83**, 1184–1186.

[11] Chang, M.N. (1990), *Weak convergence of a self-consistent estimator of the survival function with doubly censored data*, Ann. Statist. **18**, 391–404.

[12] Chang, M. N. and Yang, G. L. (1987), *Strong consistency of a nonparametric estimator of the survival function with doubly censored data*, Ann. Statist. **15**, 1536–1448.

[13] Dinse, G. E. and Lagakos, S. W. (1982), *Nonparametric estimation of lifetime and disease onset distributions from incomplete observations*, Biometrics **38**, 921–932.

[14] Efron, B. (1967), *The two-sample problem with censored data*, Proceedings of the fifth Berkeley Symposium on Mathematical Statistics and Probability, 831–853, University of California Press.

[15] Fan, J. (1988), *On the optimal rate of convergence for nonparametric deconvolution problem*, Technical Report, Department of Statistics, University of California, Berkeley.

[16] Gill, R.D. (1989), *Non- and semi-parametric maximum likelihood estimators and the von Mises method, part I,* Scand. J. Statistics **16**, 79–128.

[17] Gill, R.D. (1991), *Non- and semi-parametric maximum likelihood estimators and the von Mises method, part II,* Preprint Nr. 664, Department of Mathematics, University Utrecht.

[18] Groeneboom, P. (1987), *Asymptotics for interval censored observations,* Technical Report 87-18, Department of Mathematics, University of Amsterdam.

[19] Groeneboom, P. (1988), *Limit theorems for convex hulls,* Probability theory and related fields **79**, 327–368.

[20] Groeneboom, P. (1989), *Brownian motion with a parabolic drift and Airy functions,* Probability theory and related fields **81**, 79–109.

[21] Groeneboom, P. (1991), *Discussion on: Age-specific incidence and prevalence: a statistical perspective, by Niels Keiding,* J. R. Statist. Soc. A **154** , 400–401.

[22] Groeneboom, P. (1991), *Nonparametric maximum likelihood estimators for interval censoring and deconvolution,* Technical Report 378, Department of Statistics, Stanford University.

[23] Hájek, J. (1970), *A characterization of limiting distributions of regular estimates,* Z. Wahrscheinlichkeitstheorie verw. Gebiete **14**, 323–330.

[24] Hájek, J. (1972), *Local asymptotic minimax and admissibility in estimation,* Proc. Sixth Berkeley Symp. Math. Statist. Prob. **1**, 175–194, University of California Press, Berkeley.

[25] Has'minskii, R. Z. and Ibragimov, I. A. (1983), *On asymptotic efficiency in the presence of an infinite-dimensional nuisance parameter,* USSR-Japan Symposium, 195–229, Springer-Verlag, New York.

[26] Hoffmann-Jørgensen, J. (1984), *Stochastic Processes on Polish Spaces,* Unpublished manuscript.

[27] Jeganathan, P. (1981), *On a decomposition of the limit distribution of a sequence of estimators,* Sankhya **43**, 26–36.

[28] Jeganathan, P. (1982), *On the asymptotic theory of estimation when the limit of the log-likelihood ratios is mixed normal,* Sankhya **44**, 173–212.

[29] Jewell, N.P. (1982), *Mixtures of exponential distributions,* Ann. Statist. **10**, 1184–1186.

[30] Jörgens, K. (1982), *Linear Integral Operators,* Pitman, Boston, Transl. G. F. Roach.

[31] Karlin, S., Studden, W.J. (1966), *Tchebycheff Systems: with Applications in Analysis and Statistics,* Interscience, New York.

[32] Keiding, N. (1991) *Age-specific incidence and prevalence: a statistical perspective (with discussion)* J. R. Statist. Soc. A **154**, 371-412.

[33] Kim, J., Pollard, D. (1990), *Cube root asymptotics*, Ann. Statist. **18**, 191–219.

[34] Koshevnik, Yu. A. and Levit, B. Ya. (1976), *On a non-parametric analogue of the information matrix*, Theor. Prob. Applic. **21**, 738 – 753.

[35] Le Cam, L. (1972), *Limits of experiments*, Proc. Sixth Berkeley Symp. Math. Statist. Prob. **1**, 245–261, University of California Press, Berkeley.

[36] Le Cam, L. (1986), *Asymptotic Methods in Statistical Decision Theory*, Springer, New York.

[37] Le Cam, L. and Yang, G. L. (1988), *On the preservation of local asymptotic normality under information loss*, Ann. Statist. **16**, 483–520.

[38] Levit, B. Ya. (1978), *Infinite-dimensional informational inequalities*, Theory Prob. Appl. **23**, 371–377.

[39] Millar, P. W. (1979), *Asymptotic minimax theorems for the sample distribution function*, Z. Wahrsch. verw. Geb. **48**, 233–252.

[40] Millar, P. W. (1983), *The minimax principle in asymptotic statistical theory. École d'Été de Probabilités de St. Flour XI-1981*, Lecture Notes in Math. **876**, 76–262.

[41] Millar, P. W. (1985), *Nonparametric applications of an infinite dimensional convolution theorem*, Z. Wahrsch. Th. verw. Gebiete **68**, 545-556.

[42] Pfanzagl, J. (with W. Wefelmeyer) (1985), *Asymptotic Expansions for General Statistical Models*, Lecture Notes in Statistics **31**, Springer Verlag, New York.

[43] Pollard, D. (1984), *Convergence of Stochastic Processes*, Springer-Verlag, New York.

[44] Pollard, D. (1989), *Asymptotics via empirical processes (with discussion)*, Statist. Sci. **4**, 341–366.

[45] Pollard, D. (1990), *Empirical Processes: Theory and Applications*, NSF-CBMS Regional Conference Series in Probability and Statistics, 2.

[46] Reed, M. and Simon, B. (1972), *Methods of Modern Mathematical Physcis I: Functional Analysis*, Academic Press, New York.

[47] Robertson,T., Wright, F.T., Dykstra, R.L. (1988), *Order Restricted Statistical Inference*, Wiley, New York.

[48] Rockafellar, R.T. (1988), *Convex Analysis*, Princeton University Press.

[49] Roussas, G. G. (1972), *Contiguity of Probability Measures*, Cambridge University Press, Cambridge.

[50] Rudin, W. (1973), *Functional Analysis*, McGraw-Hill, New York.

[51] Strasser, H. (1985), *Mathematical Theory of Statistics*, de Gruyter, Berlin.

[52] Stefanski, L.A., Carroll, R.J. (1987), *Deconvoluting kernel density estimators*, University of North Carolina Mimeo Series **1623**.

[53] Tricomi, F. G. (1957), *Integral Equations*, Wiley, Interscience Publishers, Inc., New York.

[54] Tsai, W.I., Crowley, J. (1985), *A large sample study of generalized maximum likelihood estimators from incomplete data via self-consistency*, Ann. Statist. **13**, 1317–1334.

[55] Tsai, W.I., Crowley, J. (1990), *Correction on: A large sample study of generalized maximum likelihood estimators from incomplete data via self-consistency*, Ann. Statist. **18**, 470.

[56] Turnbull, B.W. (1974), *Nonparametric estimation of a survivorship function with doubly censored data* J. Amer. Statist. Assoc. **69**, 169–173.

[57] Turnbull, B.W. (1976), *The empirical distribution function with arbitrarily grouped censored and truncated data*, J.R. Statist. Soc. B **38**, 290–295.

[58] Turnbull, B. W. and Mitchell, T. J. (1984), *Nonparametric estimation of the distribution of time to onset for specific diseases in survival/sacrifice experiments*, Biometrics **40**, 41–50.

[59] van Es, A.J. (1991), *Aspects of Nonparametric Density Estimation*, CWI Tract, **77**, CWI, Amsterdam.

[60] van de Geer, S. (1990), *Hellinger-consistency of certain nonparametric maximum likelihood estimators*, Preprint Nr. 614, Department of Mathematics, University Utrecht.

[61] van der Vaart, A. W. and Wellner, Jon A. (1991), *Prohorov and continuous mapping theorems in the Hoffmann-Jørgensen weak convergence theory, with applications to convolution and asymptotic minimax theorems*, Technical Report **157**, Department of Statistics, University of Washington, Seattle.

[62] van der Vaart, A. W. (1988), *Statistical estimation in large parameter spaces*, CWI Tract **44**, Centrum voor Wiskunde en Informatica, Amsterdam.

[63] van der Vaart, A.W. (1991), *On differentiable functionals*, Ann. Statist. **19**, 178–205.

[64] Vardi, Y. and Zhang, C.-H. (1990), *Large sample study of empirical distributions in a random-multiplicative censoring model*, Preprint, Rutgers University.

[65] Vardi, Y. (1989), *Multiplicative censoring, renewal processes, deconvolution and decreasing density: Nonparametric estimation*, Biometrika. **76**, 751–761.

[66] Wu, C.F.J. (1983), *On the convergence properties of the EM algorithm*, Ann. Statist. **11**, 95–103.

[67] Cun-Hui Zhang (1990), *Fourier methods for estimating mixing densities and distributions*, Ann. Statist. **18**, 816–831.